Textbook on Freshwater Aquaculture

Textbook on Freshwater Aquaculture

Dr. B. Ahilan, *M.F.Sc. Ph.D.,*
Professor
Department of Aquaculture
Fisheries College and Research Institute
Tamil Nadu Veterinary and Animal Sciences University
Tuticorin – 628 008
Tamil Nadu

2013
Daya Publishing House®
A Division of
Astral International Pvt. Ltd.
New Delhi – 110 002

ISBN 978-81-7035-793-3

Published by : **Daya Publishing House®**
A Division of
Astral International Pvt. Ltd.
81, Darya Ganj, Near Hindi Park,
Delhi Medical Association Road,
New Delhi - 110 002
Phone: 011-4354 9197, 2327 8134
Fax: +91-11-2324 3060
E-mail: info@astralint.com
Website: www.astralint.com

Laser Typesetting : **Classic Computer Services**
Delhi - 110 035

Printed at : **Chawla Offset Printers**
Delhi - 110 052

PRINTED IN INDIA

Tamil Nadu Veterinary and Animal Sciences University

Foreword

Fisheries play a vital role in feeding the world's population, contributing significantly to the dietary protein intake of millions of people. Various types of aquaculture form an important component within agricultural and farming systems development. These can contribute to the alleviation of food insecurity, malnutrition and poverty through the provision of food of high nutritional value, income and employment generation. The rationale of aquaculture is not limited merely to socioeconomic and marketing advantages. Irrespective of the economic or other benefits of large-scale aquaculture operations, greater emphasis is laid to small scale farming in developing countries. The different technologies available for freshwater aquaculture need to be popularized in India so as to make use of the available unused water bodies for freshwater fish culture and production. A Textbook on Freshwater Aquaculture is the need of the hour to help the students of Fisheries Science in State Agricultural

Universities and in conventional universities where fisheries is taught as a subject. Keeping these in view, this book is brought out in a readable style and I strongly hope that this publication would be of immense use for different constituents of fisheries sector.

I congratulate the author of this publication for his sincere efforts.

Dr. V.K. Venkatramani, *Ph.D.*,
Director of Research and Extension (Fisheries)

Preface

Aquaculture is an important domestic provider of much needed, high quality, animal protein, generally at prices affordable to the poorer segments of society. It is also a valuable provider of employment, cash income and foreign exchange, with developing countries contributing over 90 per cent of the total global production. Aquaculture production is expected to play a crucial role in forthcoming decades in compensating for stagnant capture fisheries and in meeting increased demand for aquatic products.

Keeping the importance of freshwater aquaculture practices in the coming years, this book is prepared. This book emphasizes different fish farming practices, pre-stocking management, post-stocking management, components of the pond, carrying capacity of the pond and culture practices of commercially important freshwater aquaculture species. The intended readership is primarily undergraduate and postgraduate students of aquaculture. It is also hoped that fish farmers and entrepreneurs will also found this textbook useful.

I am indebted to Dr. V.K. Venkatramani, Ph.D., Director of Research and Extension (Fisheries), Tamil Nadu Veterinary and Animal Sciences University, Thoothukudi for his Foreword. I am thankful to the Dean, Fisheries College and Research Institute, Professor and Head, Department of Aquaculture, Fisheries College and Research Institute, Thoothukudi.

B. Ahilan

Contents

Chapter 1

Introduction

Aquaculture and agriculture are not strictly parallel developments in food production, even though food gathering, hunting and fishing might have started at about the same time in human history. The complexity of aquaculture as a multi-disciplinary activity, even more complex than agriculture, is perhaps one of the reasons for the late start of modern aquaculture. Asia, in particular China, has dominated aquaculture developments in the world right through history and still China dominates the world scene by accounting for over 65 per cent of global aquaculture production. Though India is second only to China, it accounts for only 7 per cent of global aquaculture production. Aquaculture is one of the fastest growing food producing sectors of the world and has achieved a reputation as a significant contributor to poverty alleviation, food security and income generation. Aquaculture is an important domestic provider of much needed, high quality, animal protein generally at prices affordable by the poorer segments of society. It is also a valuable provider of employment, cash income and foreign exchange, with developing countries contributing over 90 per cent of the total global production.

The quantum of fish currently exploited by capture methods from our natural water bodies is inadequate to meet the demand of the ever increasing population and to combat malnutrition. India need more than 12 million tones of fishes per year but our present production is 5 million tones. In this context the importance of Aquaculture is greatly felt as that of Agriculture and Animal husbandry. It could be an alternative venture to offer protein rich food and Income.

Next to agriculture, aquaculture had been a lucrative enterprise in places like Kerala and West Bengal. Good monsoons and extensive freshwater and brackish water resources promise a good fortune for aquaculture in the country. These water resources together with the technical manpower can be elegantly manipulated for the production of cheap, but high quality protein for the mankind.

Inland Fish Production

As against 0.24 million tones of fish produced in 1950-51, the production of inland fish in the country during 2003-04 was at 3.4 million tones, and this increase in the fish production has placed the country on second largest producer of inland fish.

Inland Water Resources

The river system of the country comprises 14 major rivers (catchments >20,000 km^2), 44 medium rivers (catchments 2,000 to 20,000 km^2) and innumerable small rivers and desert streams. Different river systems of the country, having a combined length of 29,000 km, provide one of the richest fish genetic resources in the world. Besides, the resources under ponds and tanks have been estimated at 2.254 million ha and in the coastal area 1.2 million ha has been identified as potential resource for finfish and shellfish farming.

Riverine Resources of India

The inland water resources harbour the original germplasm of one of the richest and diversified fish fauna of the world, comprising 930 fish species belonging to 326 genera, out of 25,000 total fish species recorded world-wide. The major river systems of India on the basis of drainage can be divided broadly into two–(*i*) Himalayan river system (Ganga, Indus and Brahmaputra), and (*ii*) Peninsular river system (East coast and West coast river system).

Reservoir Resources

Reservoirs constitute the single largest inland fishery resource, both in terms of resource size and productive potential. A detailed study made by the FAO in 1995 has estimated a total of 19,370 reservoirs in the country with a total area of 3.1 million ha.

The reservoirs in the country are distributed under divergent geoclimatic, morphometric and edaphic environments. Reservoirs are classified generally as small (<1,000 ha), medium (1,000–5,000 ha) and large (>5,000 ha). The details of distribution of small, medium and large reservoirs in India are given in the Table 1.

Table 1: Distribution of Different Reservoirs in India

	Small	*Medium*	*Large*	*Total*
Number	19,134	180	56	19,370
Area (ha)	1,485,557	527,541	1,140,268	3,153,366

Fish production from different categories of reservoirs like small, medium and large has been estimated at about 50 kg/ha/year, 12.3 kg/ha/year and 11.5 kg/ha/year, respectively, the average being about 20 kg/ha/year. This production is very low in comparison to countries such as Thailand (65 kg/ha/year), Russia (88 kg/ha/year) and Sri Lanka (100 kg/ha/year). The present low levels of production from the reservoirs is on accounts of many reasons such as lack of fish seed production and stocking, inappropriate gears and crafts, poor landings and marketing channels, absence of closed-season and other inadequate management measures like weak cooperatives, stranglehold of middle-men, etc.

Coldwater Fisheries

The ecological conditions of the fishery resources of the upland areas namely coldwater streams, brooks, rivers, natural and man-made lakes are quite different from that of the plains. There are 2 types of streams namely snow-fed and spring-fed. These streams are characterized by rocky and gravelly bottoms, high transparency, low temperature, high oxygen level due to high velocity and continuous water flow and their primary and secondary productivity are low. Temperature plays a crucial role in the occurrence and distribution of fishes in these areas. The common coldwater fishes are given in Table 2.

Freshwater Aquaculture

A production of 600 tonnes/ha/year of tilapia in cages has been recorded in Laguna Bay in Philippines. At Mannheim, Germany, carp production system yields more than 150 tonnes/ha/year in two crops. Thailand farmers have obtained the catfish (*Clarias*) production at the rate of

Table 2: Common Fishes of Upland Areas of the Himalayas and Peninsular Plateau

Group	*Family/Sub-family*	*Fish Species*
		Indigenous fish stocks
Mahseer	Cyprinidae/Cyprininae	*Tor putitora, T.tor, T.khudree, T.mosal, Tor malabaricus, Neolissocheilus hexagonolepis,*
Snow trout	Cyprinidae/Cyprininae	*Schizothorax plagiostomus, S.progasrus, Schizothoraichthys scurvifrons, S.niger*
Brails	Cyprinidae/Rasborinae	*Barilius bola, B.barila, B.vagra*
Minor carp	Cyprinidae/Cyprininae	*Labeo dero, L.dyocheilus*
Loaches	Cobitidae	*Botia* spp., *Nemacheilus* spp.
Catfishes	Sisoridae	*Glyptothorax pectinopterus, Glyptosternum venticulatum*
		Exotic fish stock
Trout	Salmonidae	*Salmo gairdenri (Oncorhynchus mykiss), Salmo trutta fario, Salvelinus fontinalis*
Common carp	Cyprinidae	*Cyprinus carpio communis, C. carpio specularis, C.carpio nudus*
Trench	Cyprinidae	*Tinca tinca*

Source: NRC-CWF (ICAR), Bhimtal.

208 tonnes/ha/year in three crops. Taiwan achieved more than 20 tonnes/ha/crop of tiger prawn. Uruguay and Hawaii are not far behind. In Norway, the salmon cages produce 150 tonnes in two-year cycles as a commodity par excellence. These technologies are highly capital intensive. However, the average productivity of freshwater fish from ponds in Israel and China has been around 15 tonnes/ha/year.

India has also witnessed considerable advancements to achieve high rates of productivity in finfish and shellfish farming systems with production up to 15 tonnes/ha/year of carps, 13 tonnes/ha/5 months of white prawn, 6-8 tonnes/ha crop of tiger prawn, 54 tonnes/ha/5 months of catfish and 6 tonnes/ha/year of rainbow trout. Similarly, high production was achieved for the exotic channel catfish culture in Tanjavur through semi-intensive aquaculture technologies following a high stocking density, aeration, balanced feed and management of water quality by HLL.

With the average aquaculture production being around 2.2 million tones in India, the sector recorded 6 per cent annual growth rate. Carps, catfishes, prawns and mussels form important components of culture practices in terms of cultivable species, being diverse to suit the varied ecological conditions of different water bodies, also meet the regional preferences. Being basically a carp country, both indigenous and exotic carps, catla, rohu, mrigal, silver carp, grass carp and common carp account for the bulk of the production. Several other candidate species include *Labeo calbasu, L.gonius, L.bata, L.dussumieri, L.fimbriatus, Barbodes carnaticus, Puntius pulchellus, P.kolus, P.sarana, Cirrhinus cirrhosa* and even the minor carps fetch high prices.

Three Indian major carps contribute to about 84 per cent of India's aquaculture production, which is mostly rural based. Increasing the unit area production of these 3 carps as well as the exotic carps, and expanding the culture area is also possible through integrated fish farming as a package can lead to substantial increase in freshwater fish production. Similar thrust is also required to give a boost to aquaculture in the upland areas. India's carp production is also mainly rural but integration of fish culture in the rural development scene in India is yet to be effective. The seed production of carps has been standardized, with commercial hatcheries producing over 19,000 million fry annually.

Water Resources

The freshwater aquaculture resources in the country comprise 2.25 million ha of ponds and tanks, 1.3 million ha of beels and derelict waters, 2.09 million ha of lakes and reservoirs as also 0.12 million km of irrigation canals and channels and 2.3 million ha of paddy fields, a portion of which is amenable for fish culture. Considering the availability of water area exclusively in the form of ponds and tanks, it is observed that only 45 per cent of area has been brought under culture. This shows the potentiality of horizontal expansion of the sector in the years to come.

Fish Production

The total inland fish production in the country registered growth from 1.2 tonnes in 1986 to 3.5 million tones in 2004-05, in which the contribution of freshwater aquaculture was way ahead of either brackish water aquaculture or capture fisheries sector on a constant basis. The present freshwater aquaculture production accounts

Table 3: Fish Production Range Under Different Culture Systems

Culture System	*Average Production Rates (tones/ha/year)*
Composite fish culture	4–6
Intensive pond culture with feeding and aeration	10–15
Clarias culture	3–5
Sewage-fed fish culture	3–5
Integrated fish farming with poultry, pigs and ducks	3–5
Pen culture	4–5
Freshwater prawn culture	1–2
Shrimp farming	2–5
Running water fish culture	25–50 kg/m^3
Cage culture	10–15
Biogas slurry-based system	3–5
Aquatic weed-based culture	3–4

for over 70 per cent of total inland fish production. The production trends show a consistent APR of 5 to 6 per cent, being significantly higher than other sectors. Carps, both Indian and exotic, contribute to over 90 per cent of the total freshwater aquaculture production. This sector alone is contributing to about one-third of the total fish production in the country. Flexibility in area of operation and scale of investments as also compatibility of freshwater aquaculture practices with other farming systems coupled with high potentials of eco-restoration in terms of organic recycling and waste treatment have made freshwater aquaculture a fast-growing farming activity in the country. Fish production range under different culture systems are given in the Table 3.

Chapter 2

Components of an Earthen Aquaculture Pond

Pond is an embankment covered by bunds to hold the water inside.

Pond normally comprises of:

1. Bunds
2. Canals - Feeder Canal and Drain Canal
3. Inlet
4. Outlet
5. Harvesting pit

Bund

The earthen enclosures covering all the sides of the pond to retain water inside are called bunds. The two sides of the bund are called slopes. The flat portion of the bottom of the slope just above the pond bottom is called bench or berm which is constructed to prevent damaging or sinking of the bund.

Canals

Every pond will have two canals a feeder canal in the

form of PVC pipe and a drain canal. The feeder canal supplies water to the pond through inlet. The drain canal will run along the side of the pond at one side in to which the water will be drained through an outlet. The drain canal will be placed at a depth, below the level of the pond bottom to facilitate easy gravitational flow of the drained water.

Inlet

The inlet is a pipe extending into the pond from the feeder canal. It is normally provided with a filter bag at its open and inside the pond to prevent the entry of unwanted fish.

Outlet

The outlet is placed at one end of the pond towards which the pond bottom has a gradient.

Harvesting Pit

A pit like structure measuring 50–100m^2 and located near the outlet of the pond is called harvesting pit.

Types of Fish Ponds

Fish ponds are of various types:

1. Nursery Pond
2. Rearing Pond
3. Production Pond
4. Breeding Pond

Nursery

It is smaller (12 × 6 × 1m) and it is mainly used to nurse the hatchlings for a period of 2 or 3 weeks. The depth of the pond is 1–1.5 m.

Rearing

It is fairly larger and has a size of 25 × 12 × 1m. The fry are grown for about 2 months until they attain fingerling stage (4–10 cm) Depth 1.5–2m.

Production Pond

Fingerlings are raised to marketable size fish. The size of this pond varies from 0.1–2.0 ha as ponds larger than 2 ha are unsuitable for efficient management.

Wintering Ponds

In regions where the winters are long and hard, adequate measures must be taken to protect the carp during that period. Wintering pond must be located near the farm building to ensure that they can be properly watched.

Utility

To protect the fish against those winter dangers likely to affect that flat shallow ponds used for summer growth. These can freeze over in winter.

The dangers are:

(*a*) Frostbite by contact with ice

(*b*) Asphyxiation under the ice

(*c*) Danger of increased acidity in the water of regions with low alkalinity through winter rains and melting snow.

Wintering pond should be deep and the greater the threat of icing the deeper they should be. Depth near the monk should be at least between 1.50 to 2 m.

The ranges of wintering ponds can be relatively limited from 0.5 to 3 ha. It is better to have small wintering ponds in good condition than large ponds which are more or less silted up and filled with grass. It is best to have wintering ponds for each species of fish. Throughout the summer these ponds should be kept dry. The bottom of wintering ponds should be rather hard but not excessively so. Wintering ponds should not be used as summer growing ponds.

Storage and Holding Ponds

These are intended to hold carp for eating from the time the growing pond is emptied in the autumn until the fish are sold. Storage ponds are small and built of earth.

Quarantine Ponds

Small ponds (0.5 ha or less) can be generally treated from the bank. Chemicals are usually dissolved or diluted in water and dispersed as evenly as possible over the surface of the pond by broadcasting in some manner. Buckets, hand operated sprayers have been effectively utilized to treat small ponds.

Characteristics of a Fish Culture Pond

Location

Select land with a gentle slope and lay out ponds to take advantage of existing terrain.

Construction

Ponds may be dug into the ground, they may be partially above and partially within the ground, or they may be below original ground elevation, side slopes and bottom should be well packed during construction to retard erosion and seepage, soil should contain a minimum of 25 per cent clay; rocks, grass, branches, and other foreign objects should be eliminated from levees.

Core Trenches

A core trench should be dug beneath the perimeter levee of the facility, and it is desirable to provide a core trench under each pond levee to prevent lateral water seepage.

Pond Depth

Depth should be 0.5 to 1.0 m at the shallow end, sloping to 1.5 to 2.0 m at the drain end; deeper ponds may be required in northern regions where the threat of winterkill below deep ice-cover exists.

Pond Configuration

Ponds should be rectangular or square.

Drain

A drain means of rapidly draining the pond should be provided, draining should be completed within 3 days.

Inflow Lines

Inflow pipes should be sufficient capacity to fill the pond within 3 days. If surface water is used, the incoming water should be filtered to remove undesirable organisms and debris.

Total Water Volume

Sufficient water should be available to fill all ponds on the facility within a few weeks and to maintain them full throughout the growing season; sufficient water should be available to provide 100 to 250 l/min/ha of total pond area.

Levees

Construct ponds with 2 : 1 or 3 : 1 slopes on all sides; levees should be sufficiently wide to move; at least one side should be wide enough (2 to 3 m) to provide vehicular access; road levees should be gravel; grass should be planted on all levees.

Orientation

Orient ponds to take advantage of wind mixing, or in areas where wind causes extensive wave erosion, place the long axis of each pond at right angles to prevailing wind; use hedge or tree wind-breaks as neçessary.

Physical Quality of Soil

Soil fraction should be about 90 per cent, stone and gravel not exceeding 10 per cent

Chemical Quality of Soil

pH–rear neutral

Total nitrogen > 0.1 per cent

Total PO_4 > 0.1 per cent

Organic (Carbon) >1.0 per cent

Free $CaCO_3$ > 5 per cent

Biological Productivity

Average plankton production per m^3 should range between 10 ml and 20 ml.

Pond Preparation

Inadequate pond bottom condition is one of the most critical problems in fish culture practices. Even when ponds are well managed, heavy organic matter inputs, (feed, fertilizer, plankton and sediments) result in accumulation of anaerobic, reduced materials on the pond bottom.

Pond bottom conditioning begins the day of the harvest. Pond bottom flushing should be carried out immediately after harvesting so that sediments cannot solidify. In small ponds, accumulation of sludge and other waste matter can be removed by hand, using suction pumps and dredges or with, earth moving equipments after drying. Physical removal of pond sediments after harvesting is generally carried out in small ponds. In that case 5-20 cm larger of topsoil is removed using bulldozers or tractors. The evacuated material can be used to build up and maintain the dikes.

After flushing out the pond, allow the pond to dry for a minimum of 7-10 days or until the pond bottom surface can support the weight of a person without settling, and the soil surface has opened in deep cracks. Drying the pond bottom facilitates:

1. Exposing and aerating surface sediments, oxidizing reduced compounds such as H_2SO_4, NO_2, NH_3, Fe, Methane etc., which are toxic to the cultivable fishes.
2. Decomposition and mineralization of organic matter by soil microbes.
3. Reduction in BOD
4. Disinfection of the pond bottom and desiccating egg, larval and adult stages of predators.
5. Ploughing/tilling of the pond bottom to expose and aerate underlying sediment layers that may be anaerobic.
6. Elimination of undesirable algal mats of filamentous algae.

After drying the pond bottom, it should be ploughed or tilled to a depth of 5-15 cm using a disk or plow. Anaerobic pond bottoms are generally characterized by deep black sediments, resulting from the reduction of ferric iron to ferrous iron. A hydrogen sulfide, odour indicates, more advanced anaerobic conditions in the sediment (Reduction sulfates to sulfide).

Following inspection of the pond bottom, agricultural lime stone ($CaCO_3$) should be applied if soil pH readings are less then 7.0. Liming pond bottoms serves to neutralize organic acids produced by incomplete decomposition of organic matter in reduced sediment. Lime also enhances decomposition of organic matter in acidic soils. Lime should be applied when the soil is slightly moist but not dry.

Chapter 3
Eradication of Predatory and Weed Fishes and Aquatic Insects

There are numerous predators which prey on smaller fish than on larger fishes. It is customary to note that, a culture fish during its life history is liable for predation at every stage. Predators are either vertebrates or invertebrates. However, among vertebrates most important groups are catfishes belonging to siluriformes.

Besides predators in unmanaged ponds, some varieties of uneconomic small fishes that naturally occur or accidentally introduced along with riverine carp spawn are termed weed fishes. Both these weed fishes and predatory fishes compete for food and space among the cultivable varieties of carps. Certain fishes are known to feed directly on carp hatchlings. That too, certain predatory fishes breed prior to the breeding of major carps in pond waters and go on feeding the available planktons with fast increasing in growth and size. When carp spawns are introduced, the predators are large enough to feed on them. Weed fishes have relatively good fecundity and attain sexual maturity in summer and breed even without rain prior to the monsoon. So their young ones are abundant in number during monsoon.

Table 4: Examples of Predatory and Weed Fishes

Sl.No.	*Predatory Fishes*	*Weed Fishes*
1.	*Glossogobius giuris*	*Ambassis ranga*
2.	*Wallago attu*	*Aplocheilus lineatum*
3.	*Clarias batrachus*	*A. panchax*
4.	*Heteropneustes fossilis*	*Barilius barila*
5.	*Mystus seenghala*	*Esomus danricus*
6.	*M. aor*	*Puntius ticto*
7.	*Notopterus chitala*	*P. conchonius*
8.	*Channa striatus*	*Rasbora daniconius*
9.	*C. marulius*	*Xenentodon concila*
10.	*C. stewarti*	
11.	*C. punctata*	
12.	*Anabas testudineus*	

Role of Predatory and Weed Fishes in Ponds

Predatory fishes especially the piscivorous varieties feed on the minnows and cultivable species of fishes. Minnows being prolific breeders and profuse feeders on planktonic masses, they consume almost all natural fish food organisms and occupy quite good space in the system. Therefore, these minnows not only compete for food but also compete for space with the cultivable varieties of fishes. This reason necessitates complete removal or the control of predatory and weeds fishes in carp culture tanks. Besides this, the stronger and larger sized fishes show cannibalistic behaviour and may feed on the weaker and smaller fishes. Hence, it is desirable to completely eradicate the predatory and weed fishes from ponds through management. Second reason necessitates about the consumer's preference on minnows and predatory fishes which are very low than the carp varieties. Therefore, carp culture is always preferable over predatory and weed fish culture due to above reason.

Methods of Eradication

Dewatering and Desilting

Complete dewatering of the pond followed by the drying of soil bottom is suggested as a successful method in eradication of unwanted fishes. It is economical only in small sized ponds. In heavily silted ponds, desilting after dewatering is advisable. Certain species of fishes like murrels, catfishes and anabas remain alive even in deep muds. Hence desilting helps in removing these fishes. Usually dewatering is suggested in summer, when the depth of water in ponds goes down to the minimum. This method is very convenient and economically applied in the seasonal or short perennial nursery and rearing ponds.

Netting Operations

It is rather a primitive one and not a sure method. However, by repeated netting in small water area with the help of a fine meshed dragnet ensures successful removal of the predatory fishes and minnows to a large extent. For more effective netting sinkers are used with the bottom rope. For the removal of predatory fishes from the pond subsurface soil, it is suggested to stir the bottom soil by means of bamboo poles or by human being or by any other means. Dragnet should be operated for effective removal of more numbers of predatory and weed fishes. The indigenous practice adopted by the fishermen for the removal of the fishes like murrels, singhi, anabas etc is to spread over a fine meshed net over the bottom soil of the pond. Sometimes in shallow ponds, such fishes were caught by hand picking also.

Hook and Lines

Hooks with long lines are used for catching mostly the piscivorous fishes like *Wallago attu*, murrels etc. Sometimes rods and lines are used with live baits to catch such fishes. Earthworms are used as baits in some cases.

Poisoning

Poisoning is a sure method of eradicating the weed and predatory fishes from any sized water body but its use has been kept restricted because of its high toxicity effect to the aquatic and sub-aquatic animals. It affect on planktons also. Therefore, a more care is necessary in the selection of poison, dose calculation and application in carp nurseries and rearing ponds.

Chemical methods are suggested for eradication of unwanted species because dewatering, desilting and drying is difficult in perennial ponds. It is better to suggest plant originated fish poisons because they affect on their respiratory system and the fishes die usually of Asphyxia. As a result, the flesh of these fishes may be used for human consumption. Moreover, the toxicity effect of the plant fish poisons does not last for a long time as compared to chlorinated hydrocarbon or organophosphates. Especially the endrin formulated fish poison chemicals are very strong poisonous and a little high dose of the poison may affect on the human beings and other animals if the treated waters are used. Further, the toxicity effect of the chlorinated chemicals is very much prolonged in to the water and as a result the treated water can not be used for a considerable period for fish culture. On the other hand the plant originated fish poisons example, mahua oil cake aids to the fertility of the pond as a fertilizer after their application.

The fish poison selected should have the following characteristics :

1. Should be freely available and economical
2. Must be effective at lower concentration
3. Should not be harmful to livestock and human beings
4. Should not have any cumulative adverse effect in the future
5. The dead fish must be usable

The fish poisons are of three types:

1. Chlorinated hydrocarbon
2. Organophosphates
3. Plant originated fish poisons

Chlorinated Hydrocarbon

Chlorinated hydrocarbons like Aldrin ($C_{12}H_8Cl_8$), Dialdrin ($C_{12}H_8Cl_8O$), Endrin ($C_{12}H_8Cl_6O$) and Taffdrin-20 are used for the eradication of predatory and weed fishes. All the endrin formulated chemicals are very strong poisons and therefore, little doses are used for fish eradication. Its action in shallow ponds is quicker during sunny days but may have hardly any action in deeper water exceeding the depth of about 20 ft. Its action takes about 2-3 hours to affect Indian major carps and 4 to 8 hours to hardy fishes like tilapia, anabas, singhi, magur etc. Plankton are also affected, especially zooplankton but they may come up again after poison effect is over.

Aldrin @ 0.2 ppm, endrin @ 0.01 ppm and dialdrin @ 0.01 ppm were found effective in eradication of many predatory and weed fishes. The toxic effect remain for 2 weeks in 0.01 ppm and for 2 day at 0.001 ppm concentration. Taffdrin contains 20 per cent endrin as its effective poison. Taffdrin-20 at the low dose of 0.008 ppm can kill minnows, weed fishes etc. The higher dose of Taffdrin @ 0.02 ppm may kill the aquatic and sub aquatic animals within 8 hours.

Organophosphate

The organophosphates used in India are DDVP, phosphamidon, thiomatin etc. The dose of phosphamidon and thiomatin @ 3 to 30 ppm can be applied for complete eradication of predatory and weed fishes. The dose of 0.0032 to 0.5 ppm is effective to the aquatic insects.

Plant Originated Fish Poisons

There are quite good number of plants which could be used for fish poisoning. Various parts of plants such as stem, barks, seeds, leaves, roots can be used as the materials for fish poisoning. Some of the common plants are

1. *Barring-tonia acutangula*
2. *Crotton tiglium*
3. Derris root powder
4. Mahua oil cake (*Bassica latifolia*)
5. Sugar cane Jaggery

The stem bark, seeds or root of Barring-tonia acutangula at the rate range of 10 to 20 ppm are effective. However, *Crotton tiglium* is effective at the rate range of 4 to 20 ppm.

In Assam, *Milletia piscida* and *Milletia pachycapra* are used to eradicate unwanted fishes. Derris root powder containing ingredient as rotenone is found to be effective in the rate range of 1 to 10 ppm in different group of fishes. Even for hardy fishes the dose was given at the rate of 20 ppm. The poison acts on the respiratory system of fishes and ultimately causes death. The poison is very effective up to a depth of 1.5 meter on hot sunny days when the temperature is more than 25°C. In deeper water its effect reduces gradually and beyond 6 to 10 meter, its effect is insignificant. The poisoning effect of Derris root powder lasts for 4-12 days when applied at the range 4-20 ppm. In order to avoid risk factor, it should be applied one month before stocking of fish fingerlings in pond.

Bassica latifolia or *Madhuca indica* is very commonly known as Mahua. Oil cakes of these plants are available in market as Mahua oilcake. The active ingredients are mowrin or resparin or even saponin. It is recommended to be applied @ 200-250 ppm to fish ponds. It contains saponin, which on dissolving enters to the blood stream and through blood circulation to gills, buccal tissues. This results in hemolysis of RBC and cause death to fish. The sign of poison effect to fish exhibited through distress, inactive and finally lose their balance. MOC acts as manure when its poisoning effect is lost. Poisoning effect remains for 21 days from the day of application. Hence, its application should be done at least a month before stocking the carp spawn in nursery ponds.

Sugar cane Jaggary is used at 1 per cent concentration in fish ponds for eradication of unwanted aquatics. The effective ingredient is saponin.

The tea seed cake is made of the residue left, after oil has been pressed out from the seeds of 'oil tea', a sort of plants belonging to camellia family such as *Camellia sasargua, C.semiserrata.* It comprises Saponin, a hemolytic toxin. The concentration of saponin at 10 ppm may cause fish to lose their balance in 9-10 hrs or to die in 11 hours. The virulence of tea seed cake disappears quickly in higher temperature. Seed meal of tamarind, *Tamarindus indica* is used @ 1750-2000 kg/ha of water for eradication of unwanted fishes. NH_3 is also successfully employed for eradication of unwanted fishes and acts as fertilizer afterwards.

Other miscellaneous materials like calcium hypochlorite commonly called as Bleaching powder contains about 30 per cent chlorine, or Urea and Bleaching powder is applied successfully for removal of unwanted aquatics as the price of mahua oil cake has increased. Even turpentine at the rate of 250 litre/ha is applied for eradication of unwanted fishes. It affects on the gills and checks effective gas exchange through gills causing Suffocation and eventually death to the organism.

Other biological agents like Amphibians are controlled by repeated netting, traps and scoop nets. Especially, frog tadpoles are eradicated by application of Tea seed cake @ 25 ppm. Reptiles are controlled by fencing and scaring. Those inside the pond are removed by repeated netting. Various fish eating birds like king fishers, Herons and ducks are removed by scaring, trapping and even by shooting.

The most important mammal for feeding on fishes are *Lutro lutro*. Other mammals feed on fishes are Beavers and Racons. These mammals can be controlled by fencing, scaring and shooting.

Aquatic Insects and their Control

Aquatic insects constituted about 4 per cent of the total insect fauna existing in the world. Any water area is invariably explored by large number of insects either in their adult or larval stage feeds directly on carp spawn or indirectly competing for food with carp spawn. Therefore, the pond culture technique includes the control and removal of harmful insects from ponds play a very important role in increasing fry survival rate. Insects are distinguished from other aquatic arthropods by their three thoracic segments, bearing 3 pairs of legs, 2 pairs of wings in the adult stage. The piercing, cutting and sucking type of mouth parts of the harmful insects cause direct or indirect death to tiny fishes. However, common insects found in the culturable ponds being smaller in sizes, can not make any harm to rather bigger sized fishes including fingerlings and yearlings. Thus removal or control of insects in the stocking tank is not a compulsory.

Aquatic Insects

Out of 11 orders of class Insecta, 3 orders namely hemiptera, coleoptera, and odonata are relatively more common in fresh water ponds.

Order Hemiptera which includes water bugs are relatively more dangerous as they are completely aquatic in both larval and adult stages. They have very strong piercing type mandibles and most fierce in preying upon fry.

Other common species belonging to Hemiptera are *Belostoma, Lithocerus* (Giant water bug), *Nepa* (water scorpion), *Ranatra* (water stick insect), *Notonecta* (Back swimmers), *Anisopes, Geirs* (water spider) etc.

Order coleoptera includes *Cybister* (water beetles), *Dytiscus* (diving beetles), *Hydrophilus* (Scavanger beetles), *Gyrinus* (whirling beetles), etc. These beetles lead their both larval and adult life in pond. The larval stages are extensively aquatic and in adult stages, they were very often found flying away out of water. Hence difficulties arise for controlling them. The strong cutting mandible may cause death of the fry by their attack.

The common insects belonging to the order Odonata in ponds are Dragon fly and Dumsel fly. The larval forms are aquatic while adult forms are terrestrial. The sucking type of mouth-parts of nymphs causes death of fish fry etc.

Control of Predatory Insects

Between initial poisoning of pre-stocking management and to the time of stocking spawn in the pond, predatory aquatic insect multiply rapidly. Many aquatic insects fly from one pond to other pond. Hence, the pond which has been cleared off from insects is again repopulated with insects. For good survival of carp spawn, the nursery pond should be devoid of insects prior to stocking of spawns. Repeated dragging of pond with fine meshed dragnet does not eradicate completely the predatory aquatic insects. Selective insecticides which kill selectively the aquatic insects without killing the fish food organisms is therefore necessary. However, spraying oil to kill the insects, by checking their respiration is a well known principle and commonly used among leading fish culturists. Mustard oil or coconut oil along with cheap washing soap at the ratio of 56 : 18 kg/ha is a well known technique developed by CIFRI for control of predatory insects within few hours of its application. The notonecta are killed within half an hour. Higher doses however kill water boatman (family corixidae), beetle larvae and bugs. Sometimes the substitution of soap is done by Teepol-B-300, a detergent manufactured by Burmashell in the emulsion. The recomınended dose of Teepol is 560 ml with 56 kg of mustard oil.

The oil of Alexandrian laurel and water dispersible gammexan commonly known as Hertax U.P. are effective to kill insects. Gammexan only when applied @ 0.6-1 ppm, can safely eradicate insects. Pure gammaisomer of benzenehexachloride soluble in ethanol can kill insects within 6 hrs at 0.01 ppm. Carp spawn can tolerate chemicals up to a dose of 0.05 ppm. Fisheries Dept. of Govt. of Maharastra has described a dose of light speed diesel oil (1 litre), Hyoxid (0.75 ml) and water (40 ml) for every 200 m^2 of water surface for control of insects. The diesel oil at 50 litre/ha with 1/3 to 1/4 of washing soap can be applied to control insects. Kerosene oil is also practiced among pisciculturists for effective eradication of aquatic insects. To kill water centipedes, quick lime can be applied before fry stocking. A drug namely Dipterex crystal (concentration 90 per cent) may be scattered into the whole pond to create a concentration of 0.4–0.5 ppm in order to kill water centipedes in nursery tanks.

Chapter 4

Aquatic Weeds and Their Management

The menace of water weeds is reaching alarming proportions in many parts of the world. Weeds in short can be said as unwanted and undesirable plants which are adopted to grow and reproduce under aquatic conditions. The presence of these undesirable plants definitely causes threat to the productive potential of the water body. The problem of weeds in fishery waters is more acute in tropical and subtropical countries than the temperate countries, although aquatic weeds now pose a global problem.

It is believed that the presence of certain plants in small numbers is frequently desirable in ponds. Any slackness given to check them would lead to their excessive growth which reduces productivity of the water body by utilizing the nutrients or check the penetration of sunlight by shading. Heavy weed infestation restricts the living space for fish, fishing operation, gradual siltation and unwanted animals and parasites make their appearance and harm the existing fish population. Moreover aquatic weeds provide place to predatory and weed fishes which are prolific breeders and profuse feeders and over populate themselves. These factors limit the available foods to the cultivable carp species resulting into less fish production.

Apart from fishery waters, heavy growth of aquatic weeds also create serious problems in canals, irrigation channels, drainage, ditches and large water bodies, interfering with hydro-electric production, wasting water in evapotranspiration and so on. More over, the decaying masses of weeds and algae pollute the potable water and environment. Furthermore, the problem of aquatic weed is worsened by increased enrichment of natural water bodies by fertilizer run-off and the nutrients from human and agricultural wastes.

In India, over 140 species of plants have been reported to be aquatic weeds both within and around various kinds of water bodies. The states comprising West Bengal, Bihar, Orissa, Assam, Tripura and Manipur, have the maximum incidence of weed infestation ranging from 40 to 70 per cent, where as in other states, it may range between 20-50 per cent (Philipose, 1968). West Bengal fisheries in 1951 had suffered a loss of 45 million kg of fish per year due to the overgrown of water hyacinth in 1.5 lakh ha of water area. 56 species of aquatic weeds of which 9 were floating, 11 emergent, 16 submerged and 20 marginal were identified (Philipose, 1972)

Problem of weed infestation in Bangladesh, Burma, Srilanka and Indonesia are almost similar to India. In most of the African Countries, the problem of weed infestation is encountered in large water bodies like reservoirs and certain slow moving rivers.

Aquatic Weeds

Hickling (1962) divides aquatic weeds according to the softness and hardness of the stem of the plants. Pruginin and Zipshitz classify the aquatic weeds according to the type of association with the fish pond. From fisheries standpoint the most useful classification based on habit and habitat of plants are as the following.

(*a*) Floating Weeds

These are free floating with their leaves above the water surface, roots are within water but not attached with the soil, *e.g. Eichhornia, Azolla and Pistia.*

(*b*) Emergent Weeds

These are rooted in the bottom soil but have all or some of their leaves, or shoots above the water surface. This group includes *Nymphaea, Trapa, Myriophyllum, Otellia, Vallisneria* etc.

(*c*) Submerged Weeds

This group of weed can be further divided in to 2 types *i.e.* submerged weeds which are completely submerged within water but may be rooted in the bottom soil for instance *Hydrilla* and *Najas* etc. and free floating submerged weeds such as *Ceratophyllum* and *Utricularia.*

(*d*) Marginal Weeds

These grow on the margins or on the shore line of the water body. They are mostly rooted in water logged soils, *e.g., Typha* and *Phragmites.*

(*e*) Filamentous Algae

These algae form "algal mats" in and around main water body. Most prominent filamentous algae in ponds are *Spirogyra* and *Pithophora.*

(*f*) Planktonic Algae

These are planktonic and their rapid proliferation results into algal blooms. The well known examples of this group are *Microcystis* and *Anabaena*.

Whatever manner the aquatic weeds are classified the most important are the methods of control which can be formulated only after their identification, habit, adoptive capacity, modes of multiplication, period of dormancy of the reproductive stages and such other ecological aspects. Furthermore, utilization of certain Aquatic weeds, in a sense constitute a free crop of great potential value–a highly productive crop that requires no tillage, fertilizer, seed or cultivation. Aquatic plants have potential for exploitation as animal feed, human food, soil additives, fuel production and waste water treatment.

Control Measures

The measures adopted for the control of aquatic weeds can be grouped into following major categories:

(*a*) Manual and management methods
(*b*) Mechanical methods
(*c*) Chemical methods
(*d*) Biological methods
(*e*) Utilisation methods

Manual and Management Methods

Before the discovery of mechanical, chemical and biological methods of weed control, manual and management methods alone were used to clear the weeds. In manual methods, human labour is employed with or without the aid of simple implements. In developing countries, manual methods are the oldest, cheapest and frequently used for aquatic weed control. Manual measures necessitate constant vigil because of temporary effect and the weeds may even grow faster than they can be removed by hand.

The emergent and marginal weeds are removed by pulling them with hand or can be kept under check by cutting their floating leaves repeatedly. The floating weeds can be either hand picked or removed by wire, coir or nylon nets. For removal of rooted submerged weeds simple devices like hand pulled bottom rakes, bamboo poles with toothed prongs can be used.

Management Measures

Management measures include

(*a*) Construction or erecting barriers for checking the entry of floating weeds.

(*b*) Periodic draining and drying of ponds.

(*c*) Controlling the submerged weeds through turbidity. Algal blooms can be controlled temporarily by spraying liquid cow dung etc.

Mechanical Methods

Mechanical methods employed for control of aquatic weeds are expensive and require manual labour to complete follow up operations. Hence used in limited scale. Moreover, transportation, installation and operation of heavy machinery in swampy areas pose problems. However, at present a number of mechanical equipments like under water weed cutter, weed harvesters, weed harvester combine are available for physical control of the aquatic weeds. Therefore, mechanical control measures include cutting of weeds followed by secondary operation of collection and disposal. Other methods like dragging or chaining, dredging, draining, drying, ploughing, raking and burning etc. are employed.

Devices used for cutting includes Hand Scythe, Double Scythe, Jointed Scythe and weed saw. Other equipments include turbine compressor and hydraulic pressure machine to dislodge the weed and collect them. In USA an "Aquatic Scavenger" for cutting and removing submerged weeds was used.

Chemical Control Method

Some chemicals popularly known as weedicide can effectively control the noxious aquatic weeds. But some of their adverse effect on fish food organisms and the residual effect on the aquatic ecosystem has been felt with aqua culturists. A large number of chemicals which can be used as weedicides for the control of aquatics can be classified in to 6 groups.

1. Inorganic chemicals–Sodium arsenite
2. Organic chemicals–Xylene
3. Auxin type regulator–2,4-dichloropenoxy acetic acid (2,4-D)
4. Other growth regulators
5. Contact herbicide/weedicide
6. Algicides-rosin-amine D acetate.

Hundreds of weedicides are available in the market in different trade names with wide range of chemical structures and functions in their killing action. These may be sprayed over the foliage, water or in the soil where root penetrates. Some herbicides are species specific and some work well on all types of aquatics.

For chemical control of floating weeds like *Pistia* and *Salvania*, foliar spray of paraquat @ 0.02 kg/ha, 2,4-D ester (MCPA) @ 25-40 kg/ha, Mineral oils (kerosene and diesel @ 775-1100 litre/ha), kerosene and urea at the ratio of 5: 1, Diguat @ 1 kg/ha or paraquat mixed with aqueous ammonia is applied. Water hyacinth (*Eichhornia crassipes*) is chemically controlled by 2,4-D (2,4-dichlorophenoxy acetic acid) @ 4.5 to 6.7 kg/ha. The dose of chemical application is also dependant upon the intensity of growth, age and time of application, environmental factor and spraying equipment employed for the purpose. Taficide 80 (2,4-D sodium salt 80 per cent) and Simazine are also effectively used.

For chemical control of marginal weeds, foliar spray of sodium salt of 2,4-D @ 5 kg/ha, 2,2 dichloropropionic sodium @ 10-12 kg/ha, Dalapon @ 25-30 kg/ha, Amitrole @ 8 kg/ha respectively can be applied. Dalapon @ 15 kg mixed with 30 litre diesel and 1 kg detergent can effectively check these weeds.

For chemical control of emergent weeds, 2-4 D @ 1.5 kg/ha mixed with 1 per cent wetting agent can be effectively used.

For chemical control of submerged weeds, Sodium arsenite @ 5-6 ppm, Urea with range of 50-300 ppm, $CuSO_4$ or $CuSO_4$ with Ammonium Sulphate, Ammonia @ 18 ppm, Aquathol (199 technical Endothaldisodium 3,6 endoxohexa hydrophathalata) can be effectively used.

For chemical control of algal blooms and filamentous algae application of copper sulphate ($CuSO_4$) @ 0.025 to 0.05 ppm, Simazine @ 0.5–1.0 ppm is needed.

Biological Control of Aquatic weeds

The biological methods are considered as the most convenient and least expensive methods for fish culture without any need for equipment or chemicals. A number of herbivorous fish species are being tested as aquatic weed control agents. Fish that control aquatic weeds are classified as:

1. Grazers, if they eat stems and foliage like grass carp.
2. Mowers, if they devour the lower portions of aquatic plants and thus cut them down.
3. Roilers and rooters, if they stir the bottom sediments while foraging for food and preparing nests, there by creating turbidity for cutting off the light that the plants need for photosynthesis and growth or they root out the plants and destroy them. .
4. Algae feeders if they consume filamentous algae.
5. Plankton feeders, if they filter microscopic algae from the water.

Biological control through fishes *viz*., *Ctenopharyngodon idella*, *Puntius gonionotus* (Swingle, 1957), *P.javonicus* (Jhingran, 1983), *P.pulchellus* (David *et al*., 1970), *Osphronemus gorami* (Villadolid, 1936), *Tilapia mossambica*, *T.melanopleura* (Jhingran 1983) have been used.

Some of the important herbivorous fishes are Common carp (*Cyprinus carpio*), Grass carp (*Ctenopharyngodon idella*), The Golden carp (*Carassius carassius*), the gold-fish

(*Carassius auratus*), The Tawes (*Puntius javonicus*), the Nilem (*Osteochilus hasseti*), Tilapia, Gourami *Osphranemus gorami, O.olfax,* Milk-fish (*Chanos chanos*), *Puntius dobsonii,* and *P. pulchellus.*

The grass carp is a truly phytophagus fish provided with powerful pharyngeal teeth to tear and macerate plant material. Of the different species of Tilapia, *Tilapia melanopleura, T.zilli, T.rendalli, T. quineensis, T.heudeloti, T.mossambicus* and Nile tilapia (*Tilapia nilotica*), are known to control weeds. Even silver carp *Hypophthalmichthys molitrix* feeds on phytoplankton there by controlling planktonic algae.

Control by Snails

A tropical freshwater snail *Marisa cornuarietis* (Seaman and Porterfield 1961) and *Pomaca australis* have been found to feed voraciously on some submerged aquatics. The former is found in Colombia and the latter in Brazil. In India, *Pila globossa* (Thomas, 1975) has shown a promise to control *Salvinia molesta* from the waters of Kerala state.

Control by Mammals and Other animals/Birds

Aquatic rat, *Mycocaster coypus* is a large sized rodent which is a native of South America has been found to be a powerful bioagent. In Israel and Poland, these aquatic rats are introduced in to fish pond for controlling weeds like *Typha, Phragmites.* But its burrowing habits and non-availability in large numbers are the constraints in its use as bioagent. In U.S.A. adult Manatee (*Trichochus manatus*) was tried experimentally for controlling aquatic vegetation. However, the availability of the animals in adequate numbers is a problem. Marginal grasses and marsh plants are effectively controlled by grazing cattle. Cray fish (Dean, 1969) and Turtle (Shah and Tyagi, 1985) have also reported to control aquatic weeds.

Swans, Ducks and Geese feed on algae and small sized weeds like Wolffia, Lemna and marginal grasses (Ross, 1971).

Control by Insects

Several insects are known to forage upon aquatic plants. It was reported that, about 58 insect species were either phytophagus or used tissues of hydrophytes as food under certain conditions. Three species of insects *viz.* (i) *Agasicles hygrophila* (Flea Beetle), (ii) *Amynothrips andersoni* (alligator trip) and (iii) *Vogtia mallio* (Stem borer moth of Alligater weed) which are native of south America, have been very successfully employed in U.S.A. Insect enemies of water hyacinth *Neochetina bruchi* and *N.eichhorinae* (weevils), *Acigona ignitalis*, *A.infusella* (Pyralidae), *Epipagis albiguttalis* (Stem borer), *Cornops longicorne* (aquatic grass hopper), *Orthogalumna terebrantis*, *Leptogalumna* spp. (mites) and *Thrupticus* spp. (Diptera) were found effective. Larvae of *Arzama densa* feed on water hyacinth causing death of some plants and prevent seed formation in others. In India, *Gesonula punctifrons* was found to attack on water hyacinth. Nigritulus, fruit eating weevil is reported as good biocontrol agent for *Ludwigia adscendens*. The insect species *Neochatina eichhorinae* released in Florida was found for control of water hyacinth. Hussain and Jamil (1989) have reported the control of water hyacinth through insects.

Salvania (Sankaran, 1976) and Pistia (Chaudhuri and Janakiram, 1975) are also known to be predated by insects. These include *Samea multiplicaulis*, *Cryptabagous singularis*,

Paulina acuminata, Noctuid (*Namengana pecticornis*). A moth (*Paraponyx stratiotata*) and weevil (*Litodactylus leucogaster*) were effective for biological control of Myriophyllum in Yugoslavia.

Control by Fungi, Bacteria and Viruses

Attempts have been made to isolate pathogenic organisms to employ them as biological agent to control the weeds. Species of *Pythium, Curvularia, Verticillum, Aspergillus, Pencillium, Eunotium* and *Torula* were reported to control effectively Hydrilla. Fungal parasites *viz.*, *Cercosporium, Cephalosporium* species, and *Myrothecium* species are to be attacking water hyacinth. Cyanophages are being explored for the control of blue green algae.

Control by Interspecific Competitions

This is the simplest and well known method. It is practiced by fish culturists in parts of India to destroy algal blooms caused by microcystis by keeping a cover of *Lemna* (duckweed) for about a week or so. Floating weeds like water hyacinth, *Pistia, Salvania* and *Spirodella* are sometimes used to suppress the growth of submerged weeds by cutting sunlight. But this method has certain limitations, because introduction of water hyacinth is very risky and secondary transportation of weed is difficult. The method is not suitable for large water bodies, but quite suitable for manageable small ponds.

Control through Utilization

This is the new concept which has originated due to heavy costs entailed on the control of aquatics by manual, mechanical and chemical methods. The negative economic role which the aquatics play can be reversed through utilization of various weeds.

Usefulness of Aquatic weeds can be categorized as,

1. Weed as fertilizer.
2. Weed as feed for Animal/bird and fish.
3. Weed as leaf protein.
4. Weed as food crops.
5. Weed as a source of energy.
6. Weed as wastewater treatment.
7. Aquatic weeds for pulp, paper and fibre.
8. Other uses.

Algal Blooms

Algal blooms are invariably caused by unicellular and filamentous algal and may impart any colour to the water like green, yellowish brown, reddish brown, bluish green etc.

- ☆ Chlorophyceae (Green algae)
- ☆ Xanthophyceae (diatoms)
- ☆ Bacillariophyceac (diatoms)
- ☆ Chrysophyceae (golden brown flagellates)
- ☆ Cryptophyceae (biflagellates)
- ☆ Dinophyceae (dinoflagellates)
- ☆ Euglenophyceae (blue green algae)
- ☆ Myxophyceae (blue green algae)

Temporary Algal Bloom

Euglena, Chalamydomonas, Volvox, Peridinium

The appearance of these algae in high densities may be avoided by suspending the application of fertilizer and

Table 5: Eradication of Aquatic Weeds by Chemicals

Type of Weed	Chemical and Dose	Method of Application	Remarks
Blue-green algae and algal weeds	a) 1 ppm copper sulphate b) 0.5-1.0 ppm Simazine. c) 0.1-0.3 ppm Diuron (1kg/ha) d) 0.2 ppm Dequat (1kg/ha) e) 0.2 ppm Paraquat (1kg/ha)	Copper sulphate packed in cloth bags is immersed in water: applied once in week if needed for a number of weeks; other chemicals are sprayed.	Copper sulphate is economical, easy for use and safe; safe level for human 2.5 ppm and for cattle 100 ppm.
Floating weeds	a) 80% 2-4 dichlorophenoxy acetic acid (5-7 kg/ha) b) 80 per cent 2,4-D (4-6kg/ha) + small quantity of 0.1 per cent surf detergent c) Simazine (5kg/ha) d) 15 ppm ammonia e) 0.2 ppm Dequat (1kg/ha) f) 0.2 ppm Paraquat (1kg/ha)	Spraying	–
Emerged weeds	a) 80 per cent 2,4-D (4-6kg/ha) + Small quantity of 0.1 per cent surf detergent b) 5 per cent Dalapon (2-2 dichloropropionic acid) (5-10 kg/ha)	Spraying	Dalapon solution is made by diesel oil; dichlorophenyl and Fenae applied after draining the water

Contd...

Table 5–*Contd...*

Type of Weed	*Chemical and Dose*	*Method of Application*	*Remarks*
	c) Dichlorophenyl (2-6 dichlorobenzonitrile) (10kg/ha) d) Fenae (2,3,6 trichlorophenyl acetic acid) (10 kg/ha)		
Submerged weeds	a) 15 ppm ammonia b) 3-5 ppm Simazine c) 250 ppm urea d) 4 ppm sodium arsenite e) 500 ppm Superphosphate f) 40 pppm Sulphur dioxide g) 0.2 ppm Dequat (1kg/ha) h) 0.2 ppm Paraquat (1kg/ha) i) Dichlorophenyl (2-6 dichlorobenzonitrile) (10kg/ha) j) Fenae (2,3,6 trichlorophenyl acetic acid) (10kg/ha)	Spraying	a) Sodium arsenite is to be carefully handled as even a dose of less than 0.05 is unsafe for human drinking, however, fish tolerate upto 12 ppm. b) Endothal is not widely used as even doses less than 0.3 ppm are toxic to fish. c) Dichlorophenyl and Fenae are to be used after draining the water from fish pond.

Contd...

Table 5–*Contd...*

Type of Weed	*Chemical and Dose*	*Method of Application*	*Remarks*
Marginal weeds	a) 80 per cent 2,4-D (5kg/ha) + Small quantity of 0.1 per cent surf detergent b) 0.2 ppm Dequat (1kg/ha) c) 0.2 ppm YParaquat (1kg/ha) d) Dichlorophenyl (2-6 dichloro-benzonitrile) (10kg/ha) e) Fenae (2,3,6 trichlorophenyl acetic acid) (10kg/ha)	Spraying	Dichlorophenyl and Fenae are to be used after draining the water from fish pond.

supplementary feeds for some time.

Persistent Algal Bloom

Blue green algae, *Microcystis, Anebaene, Osciallatoria, Nostoc, Spirulina,* appear in profusion, it may cause a surface scum which inturn may lead to fish mortality. The penetration of light to deeper waters is affected. Excess level of N, PO_4, CO_2 and organic matter are considered responsible for the fast propagation of this species. Besides increase in surface temperature and pH>9 are also said to influence the formation of persistant algal bloom.

Effect

The cells of this bloom choke the gills of fish leading to death. Toxin will be released, depletion of oxygen due to decay.

Control

1 ppm copper sulphate solution or copper sulphate at 800g/ha packed in cloth bags and immersed in ponds is known to eradicate algal bloom, particularly of blue green algae. Sulphuric acid 125 ppm or 0.5 ppm simazine may also control microcystis bloom at 0.3 to 0.5 ppm.

Biological Method

The algal bloom can be controlled by introducing duckweed *Lemna* which forms a shade over the algae and prevent their growth. Silver carp-algal feeding fish may also be introduced.

Chapter 5

Pond Fertilization

Pond fertilization is one of the key factors in increasing the maximum carrying capacity. Farmers adopted the method of manuring to rear fry ages ago. In nursery and rearing ponds, fertilization is aimed at developing natural food and saving artificial feeds. Satapathy (1990) has described a positive relation in nutrient, Phytoplankton, the primary producer of pond, carryout photosynthesis, converting the inorganic materials into organic nourishment needed for their propagation and growth. This is a link in the food chain and food web process that occurs in the ecosystem. In a pond ecosystem, the materials are in a constant rate of circulation, mainly through food chain. This is called "pond material circulation".

The fertilizers used in fish culture ponds are of two categories (1) organic (2) inorganic.

Organic Manure

Increasing attention has been paid towards recycling of various agricultural and animal wastes through aquaculture production process for enhancing fish yield. Proper management ensuring continued maintenance and building up fertility of an ecosystem is indispensable for greater productivity. This necessitates the importance of organic and inorganic fertilizers for enhancing the productivity of an ecosystem. With the rising cost of fossil fuel an acute scarcity in the area of energy and production of chemical fertilizer is felt with the entrepreneurs. That too availability of chemical fertilizer to the farmer at reasonable cost is fast declining. There is thus a need for utilizing organic manure/wastes for supplementing chemical fertilizer in the present context. That too, the accumulation of wastes/byproducts can generate pollution to the environment. If such great amount of excreta (waste) is not disposed off, it must pollute the environment. Therefore, this great amount of wastes can be recycled for good production.

Advantages

1. The advantage of applying organic manure is to improve the soil structure, water holding capacity, drainage, base exchange capacity and also check soil erosion.
2. Due to energy crisis, prohibitive cost of chemical fertilizers and poor purchasing power of marginal and small farmers, it is necessary to use organic manure/wastes to its maximum potential with proper technology to meet the shortage of chemical fertilizer and also for improving the culture system.
3. The use of organic manure has the advantage of converting unusual surplus wastes into useful product for use in agriculture, aquaculture and also for certain livestock rising.
4. Besides, nitrogen (N), Phosphorus (P) and Potassium (K), Organic manures are a potential source of micronutrients.

Varieties of Organic Manure

Organic manures are mainly farm animal excrement. Generally the term refers to manures containing organic matter. Now a days organic manure are applied as base and additive manures to fish ponds in most part of the country although inorganic fertilizers are applied in scanty. The manures often used in fish ponds are (1) faeces and urine of livestock and poultry (2) green manures (3) compost and (4) silkworm faeces. However in china, besides these above noted organic manure, human excrement (Night soil) is used not only in agriculture but also in aquaculture practices. Only through microbial decomposition, the organic manure is converted in to nutrients that the plants can absorb.

Factors Influencing the Composition and Nutritive Value of Animal Wastes

All the animal wastes are a mixture of semisolid excreta (dung) and liquid excreta (urine). Their composition and nutritive value depends on several factors like age, breed, variety, digestibility, composition of the feed ration, physiological condition of the animal etc.

Composition of Solid Excreta (dung)

The faeces of livestock chiefly consists of undigested food and also residue from digestive fluids, waste mineral matter, worn-out cells from intestinal linings, mucus, bacteria and foreign matter such as dirt consumed along with food. The nitrogen and phosphorus of the solid animal excreta form part of the organic compound and become available to plant only after their mineralization while potassium is in a mobile form. The dung of most of the animals is rich in phosphorus and calcium.

Composition of Liquid Excreta

In comparison to dung, urine contains more water, nitrogen and potash but less phosphorus. All the nutrients in urine are in a readily soluble or easily mineralisable form. Sheep urine is richest in mineral ingredient followed by horse urine. The liquid of most of the animal is rich in nitrogen and potash.

Nature and Characteristics of Organic Manure

(*i*) Pig Manure

Pig manure includes much organic matter and other nutritional elements such as N.P.K. Pig manure is fine complete manure. Pig faeces are delicate with moisture content ranged from 70-77 per cent, containing more nitrogen than other livestock faeces (C/N; 14 : 1), making them more susceptible for rotting. The major of pig urine is nitrogen in the form of urea and it decomposes easily.

The excretory amount of a pig is generally associated with its body weight and food intake. A pig per day discharges excreta 7-20 per cent of its body weight. A pig excretes about 1000kg faeces and 1200kg of urine in the culturing period of 8 months from pigling to an adult.

(*ii*) Cattle Manure

The elements of cattle manure are similar to those of pig manure. But cattle are ruminants and the food stuffs are repeatedly masticated, making the excrement quite delicate. Cattle manure contains less nitrogen than pig manure (C/N, 25 : 1), cattle urine contains more nitrogen than pig urine in the form of hippuric acid (C_6H_5 $CONHCH_2$-COOH). Therefore, cattle excreta decompose slowly. The average daily excreta of a cow of about 500kg weight is 25kg in which the ratio of faeces and urine is about 3 : 2. However, the quantity of annual dung and urine production per individual of cow are around 5400kg and 3600kg respectively (IFF, China).

Usually the cattle dung consists of about 75-85 per cent moisture, 15-25 per cent organic matter and 2-5 per cent mineral matter. The organic matter of dung mainly comprised of 78-90 per cent of total carbohydrates (Crude fibre + nitrogen free extract), 9-18 per cent of crude protein and 2-5 per cent of ether extract.

(*iii*) Poultry Manure

Poultry manure include the faeces of chicken, ducks and goose and are rich in both organic and inorganic matters since liquid and solid excreta are excreted together resulting in no urine loss. Poultry manure ferment quickly and their nitrogen is mostly in the form of uric acid which can not be absorbed directly by plants. If the dropping come from the cages or dropping pits, superphosphate may be added at the rate of 1kg/day/100 birds. This improves the fertilizing quality and helps the control of flies and odour. Poultry manures are more effective after decomposition. Although poultry excreted about 5 per cent of its body weight per day.

With regard to natural value of poultry manure, the dry matter content of the fresh poultry excreta from an adult egg layer ranged from 20-30 per cent and it comprises 20.2 per cent ash. The organic fraction of the excreta contained 21.5 per cent crude protein, 1.9 per cent ether extract, 13.4 per cent crude fibre and 42.9 per cent nitrogen free extract (See Gaur *et al.*, 1984).

(*iv*) Goat and Sheep Manure

The droppings of sheep and goat make very good manure, which is applied in cultivated lands as common practice by millions of farmers. The chemical composition of their excreta has a dry matter content of 42-48 per cent. The organic fractions of dung component comprised 52-93 per cent crude protein, 14-19 per cent ether extract, 27.8 to 36.4 per cent crude fibre, 40-47 per cent nitrogen free extract, and 0.35-0.77 per cent ash. Sheep and goat manure contain about 3 per cent N, 1 per cent P_2O_5 and 2 per cent K_2O.

(*v*) Night Soil

This type of organic manures is used in China for terrestrial grass production in an integrated fish farming system. The composition of night soil (human excrement) is greatly dependent on the food consumed. Nitrogen is high (C/N-3 : 1) and 70-80 per cent of it is in the form of urea. This facilitates for easy absorption. On an average, an adult excretes annually 90kg faeces and 700kg urine as waste material.

Night soil to be used as manure must be fermented before application. This is equally done by storing the manure in anaerobic conditions for 2-5 weeks. The decomposition of human excreta produces ammonia. Under air tight conditions and at certain concentration, ammonia can sterilize human waste. Quick-lime (1–2 per cent) or/and formalin (0.1 per cent–0.2 per cent) are effective in killing the harmful organisms and bacteria in night soil.

Methods of Organic Manure Application

Livestock manure as a base manure is applied by heaping it at a corner of the pond or in small piles in shallow water with a sunny exposure. These decompose and spread gradually into the water. If the manure is used as an additive it is added in small quantities every 10-15 days interval in pond water.

Constraints in Application of Organic Manure

The main constraint in the use of animal wastes for aquaculture is the problem of depletion of dissolved oxygen in water. The wastes undergo decomposition through bacterial action and the process uses dissolved oxygen creating biochemical oxygen demand (BOD). So in such cases, monitoring of dissolved oxygen and B.O.D. of the pond water is absolutely essential. However, addition of freshwater with aeration facilities to some extent solves this problem.

Inorganic Fertilizers

Inorganic manures are also referred to as chemical fertilizers. According to composition, chemical fertilizers can be (1) Nitrogenous (2) Phosphorous and (3) Potash fertilizers.

Advantages

1. Exact composition of inorganic fertilizer is advantageous.
2. Mineralization is very fast giving quick effect on pond productivity.
3. Lack of pollution. As they need not to undergo any bacterial decomposition, the obnoxious gases like CH_4, NH_3, SO_2 are least possible to pollute the environment.
4. No biochemical oxygen demand (BOD) is required for chemical fertilizers, or in other words their beneficial effect on oxygen content (requiring no decomposition).
5. Used in small quantities as additive manures, hence convenient for utilization.

However, when chemical fertilizer is applied in ponds, the first link of the food chain is principally phytoplankton which is not as nutritious to zooplanktons as are bacteria. Therefore, in chemical fertilizer treated ponds, the zooplankton propagation often lags far behind than that pond treated with organic manure. The other disadvantage is that the effect of inorganic fertilizer is rather short and difficult to control the water quality. Therefore, application of chemical fertilizer alone is no way better than alone application of organic manures. But it is always advisable to use organic manure as base manure and both organic and inorganic as additive manures with greater frequencies probably 10-15 days intervals or observing the planktonic conditions on pond system.

Nitrogenous Fertilizers

Liquid ammonia, Ammonium sulphate and Urea are the nitrogenous fertilizers used in production systems. The nitrogen content in these fertilizers vary due to different manufacturers with different trade names. Nitrogen content in liquid ammonia is 12-16 per cent, in $(NH_4)_2 SO_4$, 20-21 per cent and in urea is 44-46 per cent. Aqueous ammonia is readily volatilized, hence should not be exposed to the air for a long time. This is also used in killing the unwanted predators during pre-stocking management operation. Liquid ammonia can be applied sufficiently 7-8 days before stocking fish seed and can not be used till the culture is going on. In contrast, $(NH_4)_2 SO_4$ can be used as additive manure during culture period. However, urea does not ionize when dissolved in water. Hence cannot be absorbed directly by plants. Only after decomposed, they are ionized and made available to plants and planktons.

Phosphorous Fertilizer

Single super phosphate (SSP) or Triple super phosphate (TSP) is used as additive phosphorous fertilizer in pond culture system.

Potash Fertilizer

Potassium is also an essential nutritional element of plants. However, it is plentiful in the water as influenced by geochemical cycle of pond basin. There is no particular need to apply potash fertilizer, until and unless recommended by a technical expert.

Application Method for Inorganic Fertilizers

While applying liquid ammonia, the container containing liquid ammonia should be put under water and open the lid to let the liquid ammonia diffuse slowly. In this way, volatilization can be avoided. pH and temperature of water are to be considered while applying liquid ammonia, because in strong alkaline water, the NH_3 toxicity of water is noticed. At a water temperature of 25°C, the per cent of N_2 as ammonia at various pH is as follows:

Table 6

pH	*N_2 as NH_3 in (per cent)*
6	0.05 per cent
7	0.49 per cent
8	4.7 per cent
9	32.9 per cent
10	83.1 per cent
11	98 per cent

Source: Asian-Pacific Regional Research and Training Centre for IFF China (1987).

Ammonia at 0.3-0.4 mg/l of water is fatal to the juveniles of rainbow trout. Although silver, grass and common carp can tolerate ammonia concentration to some extent, but the maximum NH_3 concentration permitted for fish farming is 0.1 mg/l. Therefore, special care is needed when liquid NH_3 is applied as nitrogenous fertilizer.

The application amount of nitrogenous fertilizer depends on its nitrogen content. For example, if it is required to apply total 20 kg of nitrogen in a specified water area throughout culture period in installments in the form of $(NH_4)_2 SO_4$, 100 kg of $(NH_4)_2 SO_4$ is required for whole year as the nitrogen content of ammonium sulphate $(NH_4)_2 SO_4$ is 20 per cent.

Most water sources lack phosphorous. Therefore, it is important to apply a phosphoric fertilizer to accelerate the reproduction of azotobacteria and complement the action of nitrogenous fertilizer. The application amount is calculated according to the phosphoric acid (P_2O_5) content of the fertilizer. The method of calculation and application is usually same as that for nitrogenous fertilizer.

Doses of Manure Application

Dose of manure application is dependent on the inherent qualities of soil base and water. Hence before deciding the doses of fertilizers, the soil and water analysis has to be carried out.

Acidic water can be made alkaline by application of lime a week prior to manuring. In highly acidic soil (pH-4-5), the lime dose is 2000kg/ha, in moderately acidic soil (5-6.5 pH), the lime dose is 1000kg/ha and near neutral (6.5-7.5 pH), the lime dose is 500kg/ha and in slight alkaline soil (7.5-8.5 pH), the dose of lime application is 200kg/ha. In highly alkaline of (8.5-9.5 pH) lime application is not required.

The standard combination of N : P : K as 8 : 4 : 1 is commonly used to fertilize the pond, the dose of which is decided on the criteria of the soil as high, medium and low.

Table 7

	Available N (mg/100 gm soil)	*Available P_2O_5 (mg/100 gm soil)*
High	6-12	50-75
Medium	3-6	25-50
Low	3	25-0

The application of phosphatic and nitrogenous fertilizers may be so that the phosphate and total dissolved inorganic nitrogen in water do not surpass 0.25 and 1 PPM respectively. Application of fertilizer should be made when there is a real need to boost the production of plankton. This must be based on chemical analysis of soil and water.

To bring available Nitrogen to higher level as mentioned above in mg/100 gm soil, the $(NH_4)_2$ SO_4 @ 750 kg/ha/yr or calcium ammonium nitrate @ 750 kg/ha/yr or sodium nitrate @ 940 kg/ha/yr or urea @ 335 kg can be applied in installments. This means 150 kg nitrogen is made available to the soil through these fertilizers having respective per cent of nitrogen content. Similarly to bring available phosphorous to higher level as mentioned above *i.e.* 6 mg/100 gm soil, the SSP can be applied @ range of 375-470 kg/ha/yr and TSP @ range of 165-190 kg/ha/yr. This means 75 kg phosphorus is made available to the soil through these fertilizers having respective per cent of phosphorus content in it. The SSP available commercially has 16-20 per cent and TSP has 40-45 per cent of phosphorus.

Depending on the organic carbon in soil the amount of organic manure can be applied. If the organic carbon in soil is low (0.5 per cent), raw or compost farm yard manure should be applied @ 10,000-15,000kg/ha/yr. However, Pig dung can be applied @ 3-5 tons/ha. If poultry manure is applied as base manure, the rate may be 5000-7000 kg/ ha/yr. However, the dose is dependant to the inherent biological productivity of the system. It is always advisable to treat the dungs with lime, bleaching powder or formalin, so as to ensure that, these do not harbour any pathogenic or harmful organisms. With due fermentation, decomposition and treatment with sterilized agents, dungs are to be diluted and the supernatant should be sprayed in the pond for quick fertilization effect. If dungs will be applied directly as heaping on a corner place, silting process will go on making the pond more shallower gradually which will need reclamation after some years of fish culture practice. Besides this, the harmful pathogens harbouring in untreated dungs may cause severe O_2 depletion due to biochemical oxygen demand (BOD) and disease problem in intensive fish culture practice. Care is given that, the fertilizers are applied when pond water receives plenty of sun light, optimum dissolved oxygen, conducive, temperature, clear transparent, no wind, and adequate water level is maintained in pond. Hence, the pond water can be kept fertile, flexible and crisp with manuring and water filling. "Fertile" means that fish ponds have a fast material recycling and a great biomass of plankton, which are abundant as natural food for fish. "Flexible" means the color of pond water is always changeable. There are diurnal and monthly variations of planktonic mass imparting brown colour (due to zooplankton mass) in morning and green colour (due to phytoplankton mass) night. "Crips" means pond water is fertile but not turbid.

With the change of water colour, the water can be divided as (1) Fertile (2) Sterile (3) Water bloom and (4) Deteriorated water. With suitable management like manuring, rational application of aerators, and timely water filling, the water can be utilized for fish culture for high output.

Chapter 6

Transport of Seed Fish and Brooders

The mortality of seed fish (fry and fingerlings) which may be expected during transporation is mainly due to the depletion of dissolved oxygen and accumulation of gases like ammonia and carbon dioxide in the medium of seed fish carriers. These gases are lethal as they may reduce the oxygen carrying capacity of fish blood. For example, if the concentration of dissolved free ammonia exceeds 2 ppm, the seed fish especially that off carps may die. However, the lethal limits owing to carbon dioxide in fish depend on the level of dissolved oxygen. It has been reported that fry of more than 40 mm size may die at 15 ppm of carbon dioxide at a dissolved oxygen level of less than 1 ppm. On the other hand, such fry may die only at 200 ppm if the dissolved oxygen is around 2 ppm.

The traditional way of seedfish transport in earthen pots still exists in some parts of our country. This method, however, has several drawbacks such as (a) the earthen pots are likely to break in transit, (b) fingerlings may be injured owing to the shaking of pots, (c) possible only for short distances and short durations, and (d) frequent changes of medium (water) may result in mortality of seed fish owing to differences in water quality. Considering these, modern methods in which seed fish are transported in closed metal containers with oxygen packed medium, have now been propounded.

Techniques of Transport

Traditional Method

In this method, earthen pots of about 15 litres capacity are used for the transportation of seed fish. The pot is first filled about two-thirds with water of the spawning ground. The optimum pH and dissolved oxygen of the water to ensure maximum survival rate of seed fish should be between 7.5 and 8.5, and 4-5 ppm respectively. After filling the pot with water, about 50,000 spawn are introduced. It is advisable to condition the spawn in the hapa for about three days without feeding prior to transportation. Otherwise, the excreta may pollute the water in the earthen pot leading to death of seed fish. Earthen pots are carried manually on bamboo slings. To avoid the mortality of seed fish due to asphyxiation, water is changed once is every five hours, depending however on the number of spawn in the pot and the duration of transport.

Transport in Closed Containers

In this technique, seed fish are transported by road, rail or air in plastic bags filled with water and compressed air, packed in metal containers. Metal containers made of aluminium, galvanized iron and tin are mainly used.

Tins of 40 × 25 cm could be conveniently employed. Polythene bags measuring 65 × 45 cm are checked first by inflating the bag with little air and immersing them in a container with water. Leaking or damaged bag produces air bubbles in water. Every bag thus checked is then kept in a tin container and to about one-thirds of its capacity (8 litre) it is filled with aerated pond water. The fry or fingerlings starved for one day and well acclimated are then carefully introduced in requisite numbers into the bag. A tube form the oxygen cylinder is then allowed into the bag and the portion of the bag about 10 cm from the top is twisted and a string is kept ready for tying. The oxygen is then drawn in from the cylinder through the tube, until two-thirds of the bag is inflated or the top of the inflated bag is slightly below the top of the tin. As soon as the inflated condition is obtained, the string is tied round and the tin is closed. Such packed tins are kept in a cool place until they are transported to fish culture sites. To ensure better survival rate, the tins should be transported during evening or early morning. If the culture sites are distant, the tins in transit may be kept cooled by covering with water-soaked gunny bags.

Recent investigations made on the carp fry have shown that they could be anesthetized for transportation for ensuring better survival rate. The very purpose of anesthetizing the seed fish, which may lead to longer period of their survival is to minimize the concentration of toxic gases like ammonia and carbon dioxide in the medium by lowering the metabolic rate of the seed fish. Carbonic acid has been found to be the best anesthetic, because compared to other anesthetics such as Quinaldine and MS 222, it is not only cheap but also safe and easy to use. Seed fish have been found to survive for double the time of unanesthetised seed fish, besides ensuring better survival rate (90 per cent).

The carbonic acid anesthetization is to be done as follows. To about eight litres of water in polythene bags containing fry, 8 ml of 7 per cent sodium bicarbonate solution and 8 ml of 4 per cent sulphuric acid are added so as to produce a 500 ppm concentration of carbonic acid. Once carbonic acid is formed, the bag with anesthetized seed fish should be immediately filled with oxygen.

In order to eliminate the toxic ammonia from the medium and safeguard the seed fish from mortality, any of the following absorbents is added in the medium during transportation.

1. Synthetic amberlite resin
2. Pulverized earth
3. Permutit
4. Clinoptilolite (which when added in a concentration of 10-40 g/l is found to reduce the ammonia to about 50 per cent during 24 hours transport).

It is also found that addition of sodium phosphate which acts as a buffer at a rate of 2 g/l of the medium, may bring about a favorable pH of the medium for seed fish during transit.

Transportation of Trout Eggs

The fertilized eggs and 24 hours hardened eggs of cold-water fish, especially mahseer (*Tor* spp.) could be transported by packing them in moist cotton wool and placed in plastic bucket with an outer container. However for transshipping trout eggs, cardboard cartons of 20 × 30 × 20 cm size with inner Styrofoam lining are used. Inside these boxes, porous polyethylene bags containing about

4,000 eggs are sandwiched between and upper and lower moist sponge or cotton pad. Above the upper pad, a polyethylene bag containing about 1 kg broken ice is kept for maintaining low temperature.

Basis for Estimating Quantity of Seed Fish as a Standard

The number of seed fish to be put in a closed and oxygen packed container for transport may vary according to the type and size of the seed fish, mode of transport, duration of transport and the environmental temperature.

A simple formula to find out the number of seed fish for transportation in containers with water but without oxygen is given below.

$$N = \frac{(DO - 2) \times V}{C \times h}$$

where,

DO: Dissolved oxygen in ambient water in ppm

V: Volume of water in litre

C: Rate of oxygen consumption by individual seed fish in mg/kg/hr

h: period of transportation in hours

For example, if the water volume is 30 litres and its ambient oxygen content is 8 ppm, weight of the individual seed fish, 1 mg, rate of oxygen consumption 200 mg/kg/hr, and the period of transport six hours, the number of seed fish for transportation will be

$$N = \frac{(8 - 2) \times 30}{0.2 \times 6} = 150$$

The densities of seeds computed for transportation in eight litres of water under oxygen packing at two temperatures (the values have been derived based on the saturation of oxygen in water and oxygen consumption of fish) are given in Table 8.

Table 8: Number of Seed Fish Packed at Two Different Temperatures

Seed Fish	*25°C*			*30°C*		
	Duration of Transport, hr					
	6	*12*	*24*	*6*	*12*	*24*
Spawn (<8 mm)	12,000	6,000	3,000	10,000	5,000	2,500
Fry (8-40 mm)	600	350	175	500	300	150
Fingerlings (40-80 mm)	175	100	50	150	80	4

Transport of Breeders

Breeders have to be transported without shock and injury. This may be done using open canvas carriers (1 × 1.25 m) or in splashless closed and foam-lined containers with compressed air. It is always better to give dip baths to the breeders in any of the antiseptics or antibiotics, such as methylene blue (2 ppm), acriflavine (10 ppm), copper sulphate (0.5 ppm), potassium permanganate (3 ppm), chloromycetin (8-10 ppm) or sodium chloride (3 per cent), so as to protect them against infectious bacteria, fungi, etc. Further, before transport, the breeders have to be tranquilized using any one of the following anesthetics.

1. Quinaldine (2-methyl quinoline) (0.10-0.05%)

2. MS 222 (Tricaine methane sulphonate or m-aminobenzoate methane–sulphonate–0.1 ppm)
3. Sodium amytal (50-170 ppm)
4. Tertiary amyl alcohol (0.05 per cent)
5. Phenoxy ethanol (0.01-0.05 per cent)

Chapter 7

Cultivable Fish and Shellfish

Profitable fish culture aims at maximum production of edible fish flesh in a body of water by utilizing to an optimum extend the natural food and artificial feed which can be provided at an economically viable level. The basic factors to be considered while the selection of species are:

1. Rapid growth rate
2. Ability to use the natural food and accept artificial feed with low FCR.
3. Should have small head, high body
4. Should be hardy are resistant to disease
5. Easy to breed in ponds
6. Should be non-predaceous preferably herbivorous.
7. Should be compatible with of a species.
8. Palatable with high nutritive value
9. Should have good demand.

Catla catla (Catla)

It is the fastest growing indigenous carp in India. Body deep, stout, broad snout, conspicuous head, large upturned

Table 9: Cultivable Finfish and Shellfish Species in Freshwater Ecosystem

Groups	*Cultivable Species*	*Remarks*
	Warm water ecosystem	
Finfishes	**Indian major carps**: Catla (*Catla catla*), rohu (*Labeo rohita*), mrigal (*C. mrigala*) and calbasu (*L.calbasu*) **Exotic carps**: Grass carp (*Ctenopharyngodon idella*), silver carp (*Hypophthalmichthys molitrix*) and common carp (*Cyprinus carpio*)	India is basically a carp country and virtually freshwater farming system is concentrated mainly on carp culture. A range of technology for seed production and culture is available. The culture technology has not only brought revolutionary effect but also opened new horizon in fish culture in India.
Finfishes	**Air-breathing fishes**: Magur (*Clarias batrachus*), singhi (*Heteropneustes fossillis*), giant murrel (*Channa marulius*), striped murrel (*C.striatus*), spotted murrel (*C.punctatus*) and climbing perch (*Anabas testudineus*)	The air-breathing fishes are another popular group, which are cultured in freshwater aquaculture system. In fact these carnivorous fishes are swamp inhabitants. By virtue of the presence of aerial respiratory organ, they can tolerate adverse ecological conditions; therefore they are suitable for culture in unmanageable water bodies not suitable for carp culture.
Crustaceans	**Freshwater prawns**: Giant freshwater prawn (*Macrobrachium rosenbergii*), Indian freshwater prawn (*M.malcolmsonii*) and Ganga river prawn (*M.choprai*)	Most prized, delicious, freshwater aquafood. These prawns are fast growing species and suitable for culture along with carps. Monoculture of these species may yield about 800-1,000 kg/ha/year
Molluscs	*Lamellidens* spp., *Hyriopsis* spp.	Bivalves are important for pearl culture.

Contd...

Table 9–*Contd...*

Groups	*Cultivable Species*	*Remarks*
	Coldwater ecosystem	
Finfishes	**Indigenous species**: Golden mahseer (*Tor putitora*) and Snowtrout (*Schizothoracids*) **Exotic species**: Brown trout (*Salmo trutta fario*), rainbow trout (*Oncorhynchus mykiss*), scale carp (*Cyprinus carpio communis*) and leather carp (*C.carpio nudes*)	These fishes are most suitable species for culture in cold water of Himalayan and Peninsular uplands. In coldwater the role of water temperature is very crucial. Due to low temperature regime, the growth of fish is very slow thus for raising a table size fish 2-3 years is required. Cost of production is very high.

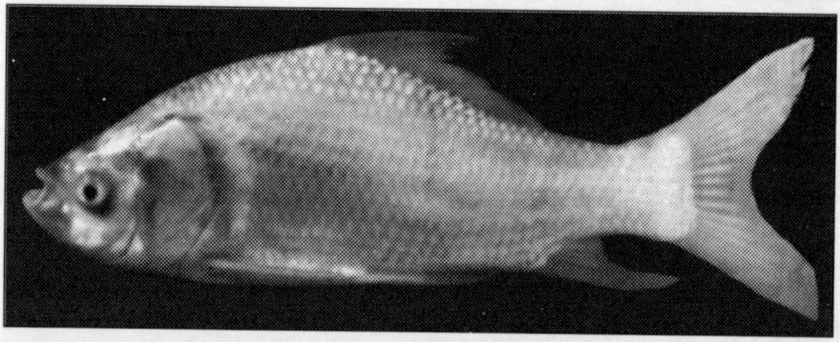

Figure 1: Catla

Figure 2: Rohu

mouth, non-fringed lips, devoid of barbells and a broad dorsal fin, back grayish colour, silvery on sides, but tends to be rather darkish in weedy waters, fins dark, scales pink.

Catla grows to over 1.5 meter length. The fish attains sexual maturity in the 2nd year and are ready to breed. It feeds on zooplankton, young ones feeds on both phyto and zooplankton. It is a fast growing species among the Indian major carps. First year growth is 35-45cm length and weighs about 1.5-2.0kg. Artificial propagation is possible widely distributed in Northern India, Pakistan, Bangladesh and Burma. Fecundity–2-4 lakhs.

Labeo rohita (Rohu)

Rohu is considered as tastiest among the Indian carps. It is distinguished by its relatively small or pointed head, almost terminal mouth with fringed cover lip. Body elongated with moderately convex abdominal. Back brownish-grey. Dull reddish scales on the sides and pink reddish fins. The body is more linear than catla. Sexual maturity is attained towards the end of the second year. Rohu grows over 91cm in length.

It is a column feeder feeds on phytoplankton, plant debris or decaying aquatic plants. The youngones feeds on zooplankton. The maximum size attained is 1m. It is fairly fast growing species. First year growth is 35-40cm and 900g. Fecundity–2.26-2.80 lakhs.

Cirrhinus mrigala (Mrigal)

Mrigal is next in importance to Catla and Rohu for culture. Linear body, small head with blunt snout, sub terminal mouth with thin non-fringed lips, dorsal fin with 12-13 branched rays. Body silvery dark grey along back, fins orange tinged with black. Length over 66cm.

Mrigal grows slower than Catla and Rohu. It attains sexual maturity in second year. It is a bottom feeder feeds on decaying organic and vegetable debris, however its young feeds on zooplankton. The maximum size attained 0.9m. Its growth in the first year is about 30cm (700g). Fecundity–2-2.5 lakhs.

Hypophthalmichthys molitrix (Silver Carp)

It is a Chinese carp introduced in 1959 by importing from Hongkong and Japan. An upturned mouth and laterally compressed body with small silvery small scales

Figure 3: Mrigal

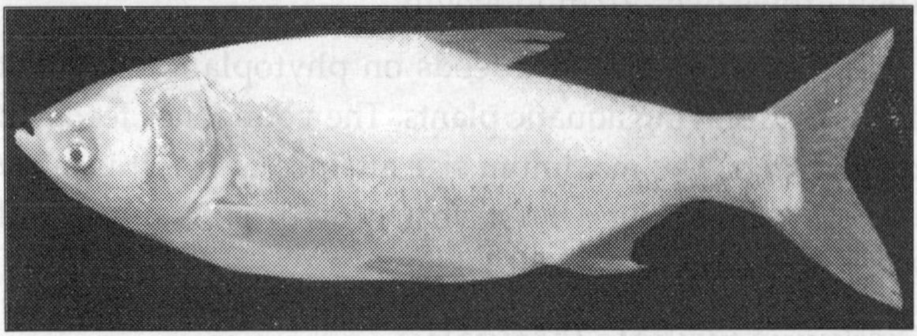

Figure 4: Silver Carp

are the identifying features of this species. Eyes, comparatively small, situated below horizontal axis of body. Gill rakers dense, interlaced, connected and covered with a spongy sieve membrane. Colour of body in live condition is silvery white, dorsally dark brown.

It is a surface feeder feeds mainly on phytoplankton of less the 0.025mm size. Young ones feed on zooplankton. Maximum size attained is 60cm. Growth in the first year is 1.5kg. Sexual maturity takes place in the second or third year. Artificial breeding was successful in 1962.

Ctenopharyngodon idella (Grass Carp)

Grass carp was introduced in India during 1959. Large sized body almost cylindrical, with flat head and round abdomen. Small eyes shorter lower jaw moderately sized

Figure 5: Grass Carp

Figure 7: Common Carp

scales with light greenish tinge. Normally it feeds on aquatic weeds and terrestrial grass (Napier). Young ones feed on zooplankton. Maximum size attained is 1.5m and 30kg. It has a very fast growth rate and maximum growth observed in the first year is 90cm and 6.9kg.

Cyprinus carpio (Common Carp)

There are lot of morphological variations through artificial breeding and natural selection of this species. Although 3 well known varieties of common carp like *Cyprinus carpio var communis* (Scale carp), *Cyprinus carpio var specularis* (Mirror carp), *Cyprinus carpio var nudus* (Leather carp).

Deep body, short compressed head, round abdomen, mouth slightly downwards with blunt snout and with two pairs of barbells on upper jaw, lower pair a little longer, long dorsal fin, thick and big scales, colour of the body in live condition varying with different living conditions usually dark grey or yellowish brown dorsally, lateral golden yellow. Apex lining of caudal fin is slightly red.

Chinese carp introduced in India during 1939 and 1957 from Srilanka and Thailand. Leather carp is confined to cold upland waters. It is an omnivorous bottom feeder, feeds on larvae of insects, worm, mollusks, leaves of submerged plants and occasioually on zooplankton. Maximum size attained is 75cm and 6kg. Fast growing fairly hardy fish. Growth in the first year is 40-45cm and 1-1.5kg.

Cat Fishes

Wallago attu (Freshwater Shark)

Snout is rounded. Mouth transverse and very large. The mouth cleft extends beyond eyes. Lower jaw is more prominent. Teeth are large, pointed and depressible. It is a largest catfish grows to a maximum size of 2m.

Mystus oar (Cat Fish)

Presence of adipose fin (maxillary barbells reaching caudal fin). Maximum size 1.5m.

Mystus seenghala

Upper jaw is longer and cleft of mouth shallow. The median longitudinal groove in head extends to the base of occipital process. Adipose dorsal fins are present. Body brownish grey, silvery on sides and abdomen. A black spot on the hind end of adipose dorsal fin. It is very much liked in North-Western states of India. Four pairs of barbells maxillary barbell reaching pelvic fin. It reaches a maximum size of 1m.

Pangasius pangasius (Pangash)

Presence of adipose fin, mouth is inferior. It reaches a maximum size of 1.5m.

Heteropneustes fossilis (Singhi)

The dorsal profile is almost straight but ventral is slightly convex. Jaws are of equal length. Lips are continuous and papillianated. There are 4 pairs of barbells, maxillary reaches to the middle of pectoral or even up to pelvic base. The dorsal fin is short and spineless. The pectoral spine is strong, osseous and pointed. The colour is blackish above and lighter beneath. Absence of adipose fin, rounded caudal fin. It reaches a maximum size of 45cm.

Clarias batrachus

It is commonly known as magur. Magur attains sexual maturity when it is one year old. It breeds in inundated waters of rice fields, swamps and marshy areas during monsoon and exhibit parental care. Sexual dimorphism is distinct as the genital papilla is pointed in male and swollen in female which gets vascularised during breeding season. Natural propagation and hypophysation technique in breeding *Clarias batrachus* is well documented in literature. Elongated first dorsal, long anal and rounded caudal. It reaches a maximum size of 40cm.

Murrels or Snake Heads

Channa marulius (Giant Snake Head)

Body is cylindrical and compressed posteriorly. Head is depressed, cleft of mouth extends beyond the eye. Colour of the back is grayish, abdomen is orange, a black ocellus encircled by whitish out line is present at upper part of the caudal fin and there are also black bands on lateral line, spots are present on posterior part of the body, dorsal fin, anal and caudal fin. The young ones have a brilliant orange coloured bands passing from the eye to the middle of head. Like other species of channa, this species is also very hardy and can stand extreme conditions of deteriorated water. It is also having great scope of culture in the derelict water. Its flesh is white in colour, good flavour and is relished by people in many parts of country. Dorsal and anal fins of this species are long and without spines. It reaches a maximum size of 1.2m.

Channa striatus (Common Murrel)

The lower jaw is longer. Cleft of mouth is oblique and hind edge of maxilla extends beyond the posterior margin of eye. There is an inner conical row of teeth on lower jaw and cardiform teeth on palate. Scales are cycloid. The lateral line is complete and curves down below 12 dorsal fin rays. Colour of back is dusky or brown and dirty white on sides and beneath. The species is very hardy and can live for about 20 hrs out of water with the help of accessory respiratory organs. It can tolerate extreme conditions of water and therefore has great potential of culture in derelict water. Stripes are present on its body. It reaches a maximum size of 90cm.

Oreochromis mossambicus (Tilapia)

It was introduced during 1952 from Africa. It breeds nearly 8 times a year.

Macrobrachium rosenbergii (Giant River Prawn)

Occurs in all river of east and west coasts of India. It is an omnivorous feeder. Maximum length 32cm (♂) 25cm (♀). Maximum weight 450g. It migrates to estuary during breeding season.

Macrobrachium malconsoni (Monsoon River Prawn)

Found in all river. Maximum length 23cm (♂) 20cm (♀).

Macrobrachium idae (Orana River Prawn)

Found in fresh and brackish water of India. Maximum length 15cm.

Identification Characters for the Fry (14-25 mm) of Indian Major Carps

Indian major carps have more than 11 number of undivided dorsal fin rays. Other morphological characters are as follows:

Catla catla

1. Large head
2. Dorsal profile convex and the ventral profile concave
3. Margin of the dorsal and caudal fins darker. No distinct spot on the caudal fin or at the caudal peduncle
4. First ray of dorsal fin black
5. Opercular region brightly reddish
6. No barbells
7. Lips thick, but not fringed

Labeo rohita

1. A dark diffused transverse band present at the caudal peduncle
2. A pair of whitish or light grayish maxillary barbells
3. Lips fringed

Cirrihinus mrigala

1. Small head and slender body
2. Triangular dark spot at the caudal peduncle
3. No barbells visible
4. Lips thin, but not fringed
5. Tip of the lower lobe of caudal fin has a reddish tinge in larger size fry

Adult

Catla catla

(Following Mishra, 1957 and Shaw and Shebbeare 1937)

1. Fin formulae : D 17-19, V 9, C 19, L.I. 40-43
2. Head broad
3. The upper lip absent, the lower moderately thick, having a continuous and free posterior margin
4. No barbells
5. Long gills, fine and closely set
6. Pharyngeal teeth plough shaped
7. Lateral line continuous to the center of the base of caudal fin, from 5½–6½ rows of scales between it and the base of the ventral fin

8. Colour: grayish above, becoming silvery on the sides and beneath

Labeo rohita

(Following Misra, 1959)

1. Fin formulae: D3/12-13; P 17; A2/5 C19; L/40-42, Ltr 6 ½/9
2. Body oblong or elongate
3. Mouth inferior, wide, transverse and protractile
4. Lip thick, fringed and with a distinct inner fold above and below
5. Snout broadly rounded
6. Pharyngeal teeth often covered with tubercles
7. Gill rakers short
8. Scales with red mark on each side, fin grayish or black
9. Lateral line from 6 to 6 ½ rows of scales between it and base of the ventral fin

Cirrhinus mrigala

(Following Misra 1959)

1. D 3/12-13, P 15, V 9, A 3/5, C 19, L/40-45, Lrt 6½–7/8 ½
2. Snout depressed, wider mouth and thinner lips
3. Colour: silvery dark to grey along the back. The pectorals, ventral and anal are orange stained with black
4. Barbells are faintly visible

5. The tips of the lower lobe of caudal fin is vermilion red
6. Eye golden green

Food and Feeding Habits of Indian Major Carps

Catla catla is a surface and mid water feeder. It is mainly omnivorous in nature and juveniles feed on aquatic and terrestrial insects, detritus and phytoplankton (Jhingran, 1966). Main food and feeding type are variable. The non-spawning adults of *Labeo rohita* are usually found feeding in littoral zones (Khan and Jhinghran, 1975). It feeds on phytoplankton and known as a column feeder. *Cirrhinus mrigala* is also bottom feeder. Juveniles at 5 cm sizes are omnivorous and adults are almost entirely herbivorous (Jhingran and Khan, 1979). Though it is benthic in nature, it migrates throughout the water column for feeding. A long intestine (10-18 times body length) is a proof of biological adoption for efficient absorption of low nutritive food.

Growth

Catla catla is the fastest growing Indian major carp. Growth of catla is reported to be better in the freshwater than in the brackishwater (Raj, 1939; Chacko 1948; 1951a). A 68 kg specimen has been recorded from Krishnagiri reservoir (Annon, 1986). In a well prepared pond, it normally attain 0.7 to 1.0 kg in first year. The marked variability in the growth rate is due to different hydro-biological and geo-climatic conditions of water bodies. The maximum size 960 mm/11.5 kg has been recorded (Khan, 1972).

The growth rate of *Labeo rohita* is relatively slower in comparison to catla. In well prepared ponds, it attains 0.7 to 1.5 kg in first year. The maximum growth is reported to be 650 mm/2.95 kg (Chacko, 1948) in a year. The maximum size 1011-1030 mm has been recorded by Khan and Jhingran (1975). It live from 13-15 years.

A great variability has been observed in the growth rate of *Cirrhinus mrigala* from different eco-systems. Chacko and Ganapati (1951c) recorded the growth ranged from 450 to 600 mm by length and 1.135 to 1.815 kg by weight in one year specimen from tanks and swamps of Godavari and Krishna Districts. The oldest mrigal has been recorded to be 1,016 mm in length and 12 years of age from river Ganga (Jhingran, 1959).

Maturity and Spawning

Catla female reared in ponds of sub-tropics is found to attain sexual maturity in the second year. Similarly, in the river Yamuna, catla was observed to mature in second year. The length at which catla attains first sexual maturity has been stated to be different by different workers 588.8 m (Chacko and Kurian, 1950), 457.2 mm (Alikunhi, 1957), 550.0 mm (Menon *et al.*, 1959) and 441.5 mm (Natarajan and Jhingran 1963).

In case of *Labeo rohita* certain percentage of both males and females reach sexual maturity in one year of age. The average size of maturity is 292-348 and 282-500 mm for males and females respectively in ponds.

C. mrigal is reported to attain its first maturity when it is about one year old. In some parts of the country ambient temperature required for growth remains low. Thus, sexual maturity is delayed by 1-2 years. Males mature earlier than the females in the same environment. However, first maturity of these carps may vary because of surrounding temperature and management practices.

Fecundity

Fecundity in catla ranges from 1.24-1.25 lakh/kg body weight (Chondar, 1970). It breeds in river during monsoon and the spawners migrate towards the littoral zones of flood plains (Jhingran, 1966). The spawning season lasts from May to August in North-Eastern India. Catla is reported to breed twice in South India during South-West and North West monsoon.

The average fecundity of rohu per kg body weight in different environments ranged from 1,09,009–5,35,000 (Khan, 1934; Alikunhi *et al.*, 1962; Chondar 1970; Khan, 1972). The number of eggs increase with the increase of size and age of fish upto 6 years and thereafter, decrease during the subsequent age (Khan, 1972). The spawning takes place once in a year during South-West monsoon during the months April/May in Assam, April-August in the Ganga river system and north Bihar (Parameswaran *et al.*, 1970; David, 1959).

The number of eggs per unit fish weight in *C.mrigala* is moderately high. The fecundity per kg body weight in mrigal from different ecosystem/climatic conditions in India ranged from 1,44,000-2,44,537 (Chondar, 1970; Chakraborty and Singh, 1963). The natural spawning of mrigal occurs in marginal areas of rivers and reservoirs where fluviatile conditions prevail.

Identifying Characters for Chinese Carps

Grass Carp

Fry

The body of the fry is elongated and somewhat slender. Both dorsal and ventral profiles almost equally straightened with relatively long head. The mouth is terminal, barbells absent and eyes moderately big. The embryonic fin-fold gradually reduced in size and persists in vestigial form. It is visible between the pectoral fin and anus at 12-18 mm length of fry, but completely disappears at 27-28 mm size.

Fingerling (40-70 mm)

Complete formation of scales is found at the stage. Vestigial median fin-fold remnant completely absorbed. Openings of lateral line canal is visible. Dorsal side of the body looks more greenish. Advanced stage of the fingerling appears identical to the adult.

Adults

The body of the adult is elongated, both the profiles gently arched and abdomen is narrowly rounded. The head is broad with short snout. Upper jaw is slightly longer and protractile than the lower jaw. Mouth is slightly subterminal and oblique. Lips are thick and simple. A labial fold runs along the upper lip. Barbels are absent. Grass carp can tolerate up to 4 per cent salinity, (Routray and Routray, 1997).

Silver Carp

Fry

The body of the fry is spindle-shaped and moderately compressed with sharp ventral profile. Mouth is slightly superior with lower jaw prominent. The small dorsal fin originates slightly behind the opposite of ventral insertion. The eyes are small and situated below the axis of the body. Generally, the colour of the body is silvery.

Fingerling

The fingerling assumes similar to the adult characters in all respect. The lateral line at the mid-lateral part of the body becomes prominent. The tubercle at the tip of the lower jaw and corresponding notch at the upper jaw become conspicuous.

Fin Formula (Jhingran, 1982)

D 3/7, A 2-2/12-4 V 1/7, L1 110-124

Adult

The body is oblong and fairly compressed. The abdomen is strongly compressed with a sharp keel from breast to vent. The head is moderately large and snout bluntly rounded. The mouth is sub-superior and the upper jaw slightly notched. The cleft is not extending to anterior margin of eyes. The eyes are small, which are visible from below the ventral surface. The diameter of the eye ranges from 5.5-8.5 mm. The gill membranes are not linked with isthmus.

Common Carp

Fry

The fry are 16-49 mm in length. The body is deep and laterally compressed, with small head and narrow protractile mouth. Two pairs of maxillary barbells are visible which becomes prominent at early fry stage. The long dorsal fin extends to opposite of mid-anal fin and the last ray of the dorsal fin is being somewhat ossified. No spot is visible on the tail or fins as seen in the Indian major carps.

Fingerling

Generally 50-100 mm sizes of fish are considered as fingerling. Acquires adult characters in morphology. The last unbranched ray of the dorsal fin gradually becomes spinous with prominent serration at its posterior edge at the advanced stage.

Fin Formulae (Alikunhi, 1966)

D 3-4/17-22, A 3/5, P 1/15-16, V 1/7-9, L 133-139, Ltr 5-6/7½

Adult

The body of common carp is elongated and moderately compressed. The abdomen is round and broad. The head is moderate and snout is blunt and rounded. The mouth is terminal and protrusible. The gape of mouth is narrow and lips are thin and continuous. The upper lip has 4 long barbells. The brachial openings are rather wide, reaching beyond hind border of opercle. Pharyngeal teeth are robust, which are molar like, broad, truncated, with crowns flattened or somewhat furrowed. They are arranged in three rows as 1, 1, 3 : 3, 1.1. The dorsal fin is long and originates opposite to ventral fin. It extends up to opposite of mid-anal fin.

Food and Feeding Habits

In its larval stages, grass carp feeds on plankton with preference to nanoplankton, rotifers and small sized cladocerans. In the later stages, they switch over to macrophytes. The fish becomes herbivore form 30 mm upwards. Young ones prefer soft plants since they often have difficulty in tearing them to suitable sized pieces, showing them relatively less phytophagus feeding habits, while adult masticates easily all kinds of food plants by the help of their powerful pharyngeal teeth (Hickling, 1967). The adults consume *Hydrilla, Vallisneria, Najas, Utricularia,* soft young leaves of *Eichhornia* and napier grass at 1.5-2.0 times of their body weight under pond culture. In Indian environment the advanced fry and fingerlings accept *Wolffia* and can eat *Lemna* and *Spirodella*. Advanced fingerlings and juveniles relish and effectively control *Azolla* and *Salvinia*.

Silver carp is phytophagus by nature. Fry feed on zooplankton, such as rotifers and nauplius of copepods. The advanced fry over 30 mm in length feed mainly on phytoplankton and secondly on zooplankton. The young and the adult exhibit almost similar feeding habits. They mainly depends on *Spirulina, Oscillatoria, Chlorella, Microcystis, Desmidium, Fragillaria, Diatoma, Ceratium, Peridonicum, Volvox, Botryococcus, Pediastrum, Spirogyra Microcystis* and *Aphanocapsa*, etc. Similar food items have been reported from gut content of silver carp from Govindsagar reservoir, (Kaushal *et al.*, 1980).

Common carp is polyphagus or omnivorous (Alikunhi, 1966) with wide food spectrum. It can adjust the feeding habit according to the local availability of natural food. It feeds on bottom flora and fauna. Particularly in the afternoon and evening hours, it vigorously stirs up the bottom in search of worms and other benthic fauna as well as the marginal area of the shallow pond. Common carp can be able to perforate silty bottom up to a depth of 12 cm (Nikolsky, 1963) and hard clay substrate up to 6 cm

(Alikunhi, 1966). Larvae and fry are exclusively zooplankton feeders. Fingerlings gradually switch over to the column and bottom dwelling food as they grow.

Growth

Grass carp exhibits an intrinsic growth pattern. The growth rate under optimal conditions appears to be normally greater than the other cultivable carps. In the first year it can grow up to 1 kg and thereafter 2-3 kg/year in temperate regions and up to 4.5 kg/yr in tropics. The growth rates of 10-28.4 g/day have been reported by many workers. (Hickling, 1960; Alikunhi and Sukumaran, 1964; Sinha and Gupta, 1975).

The growth rate of silver carp is slow in first year, but becomes faster from second year onwards depending upon the availability of food stuff. Under normal conditions the growth rate in tropical environment is 3-5 times of that in the European countries. In Indian condition its growth is 600-800 g in 1st year and 1.5–2.0 kg in 2nd year.

The growth rate of common carp greatly differs under different management practices. The growth of 600 to 800 g in first year and 1500 to 2000 g in the second year has been recorded in India.

Maturity

Grass carp is heterosexual. The Gonado somatic index (GSI) of grass carp is relatively low when compared with IMC. The GSI values of grass carp is between 8-20 per cent, average being 13.4 per cent (Alikunhi *et al.,* 1963). The fecundity is found to increase with length, weight and age of fish. The number of eggs ranges from ten thousand to 2 million with an average of 0.5 million for 5-7 kg brood.

In pond condition the development of gonads in silver carp as well as grass carp are exactly similar to IMC. The relative fecundity of pond reared hypophysed silver carp has been estimated at 1, 13, 465- 1, 47, 280 eggs with an average of 1, 30, 800 eggs per kg of fish (Chondar 1986).

In tropics, the mature male and female common carp are found to occur throughout the year, but availability becomes maximum during January-March and July-August quarters. Males maintain their oozing condition throughout the year after first maturation with regular feeding. In South East Asian countries, both males and females mature in the first year when they are 15-20 cm in total length and 80-130 g in weight.

Chapter 8

Carrying Capacity of a Pond

Total fish population that a given pond can support depends on its nourishing power in terms of food contained in the body of water or the artificial food that can be supplied to fish. As the carrying capacity of the pond is governed by various factors like physico–chemical and biological conditions, no two ponds can be stocked with equal number of fry or fingerlings even though they may be of same size and located adjacent to each other. Certain visual indices of productivity can be used to give an idea of stocking rate for a pond.

With the growth of fish, food and space requirements increase, so correspondingly larger area is required by them after sometime. Over stocking can be regulated by thinning out as the fishes grow. Very small fry require shallow warm water with plenty of zooplankton. Fingerlings and adult fish not only require larger and deeper waters to roam about but also their food requirements are different in quality and quantity then the fry stage.

Density in Nursery Pond

Spawn released in nursery pond are a week old measures about 4–7 mm in length. Nursery ponds are heavily stocked. The maximum rate of stocking in nursery is 6–10 million spawn per hectare. The rate of survival is usually low and fluctuating.

Density in Rearing Pond

Advance fry measuring 20–30 mm are stocked at the rate of 1.5 lakh fry/ha in the rearing pond.

Density of Stocking Ponds

Stocking of appropriate quality and quantity of fish seed is prerequisite for fish farming.

Table 10

Area of Water (ha)	*Stocking Density*
Up to 2	6000
2–10	5000
10–200	2500
200 and above	1000

The Carrying capacity of a pond which is reached when the maximum density of the fish population in a pond is obtained in terms of weight. At this stage the total weight of the fish in the pond will not increase further. The capacity of the pond in linked directly to the species of the fish which are raised and the artificial food distributed.

Maximum Standing Crop

The maximum weight of fish which a pond can support without gain or loss of weight taking into account the artificial or natural food available. Stocking of ponds seeks to determine the optimum quantity of fish to be released in such a way as to obtain the highest quantitative and qualitative productions under the most economic conditions. Stocking depends on the size and productivity of the ponds. High stocking with respect to the number of fish results in a better use of natural food then reduced stocking. The number of fish seeking food being higher in the first case. This leads to a higher quantitative production. Stocking in large number results in an increase in total production and a reduction in the weight and size of the fish produced. Reduced stocking results in an increase in individual growth of fish.

Stocking

The stocking density of carps for a culture pond is predetermined based on the targeted production, pond productivity and carrying capacity, species to be cultured, their feed-conversion efficiencies, size at stocking, growing period and level of management. Though fingerlings of more than 100 mm constitute best stocking material for grow-out culture, constraints in their availability often force farmers to resort to stoking of smaller fry. In intensive culture, a size of 50-100 g is preferred for stocking to realize higher survival and better growth. Generally a density of 5,000-10,000 fingerlings/ha is kept as a standard stocking rate in carp polyculture for a production target of 3-5 tonnes/ha/year. In seasonal ponds or where water level becomes limiting during summers, it is reduced to 2,000-3,000 fingerlings/ha to obtain higher growth. Although major carps are expected to reach an average of 0.8-1.0 kg in the first year, the growth rate is invariably reduced at higher stocking densities. With provision of water exchange and aeration, higher targeted fish production levels of 10-

15 tonnes/ha/year are achieved by resorting of stocking ponds at a density of 15,000-25,000/ha.

Selection of compatible species and stocking them at appropriate ratio suiting to culture environment is another important criteria in carp culture. Manipulation of species ratio is done for minimizing the interspecific and intraspecific competition for available food at various trophic levels in a pond. More than one species occupying different niches could be utilized in a pond for exploiting available natural food. Several combinations of Indian major carps only and in combination with exotic carps have been experimented under different stocking densities keeping growth as an index of performance. A combination of catla, silver carp, rohu, grass carp, mrigal and common carp fulfils the criteria of species, and has proven an ideal combination for freshwater carp culture in India. A proportion of 30-40 per cent surface feeders (silver carp and catla), 30-35 per cent mid-water feeders (rohu) and 30-40 per cent bottom feeders (common carp and mrigal) is commonly adopted depending on the productivity of the pond. When there is a possibility of providing aquatic vegetation as food for grass carp on a regular basis, the same may be included at 5-10 per cent of the total. While ponds with higher organic matter can be stocked with higher proportion of bottom feeders, surface feeder may be kept more in newly excavated ponds. Similarly, rohu may be stocked at higher ratio in ponds with higher water depth.

Routine Pond Management

- ✰ Feeding and fertilizing as necessary.
- ✰ Keeping the pond in good condition.
- ✰ Watching the trouble and disease.

Daily Management

- ☆ Check the pond for leaks
- ☆ Clean filters
- ☆ Watch fish behaviour near the feeding area.
- ☆ Feed the fish
- ☆ Add fertilizer if necessary
- ☆ Watch for predators.

Monthly Management

- ☆ Check the pond walls cut the grasses which is too long or plant more if necessary.
- ☆ Check the pond bottom. If there is too much build up of silt and organic matter, scoop this material out.
- ☆ Check for any weeds or other growth which might be a problem at harvest time.
- ☆ Check for leak in the inlet and outlet.
- ☆ Check the fertility and visibility of the water.
- ☆ Sampling the fishes once in a fortnight. Check the fishes for the signs of any diseases.
- ☆ Add lime if needed.
- ☆ Check for water quality.

Chapter 9

Water Quality Management

For optimum growth and the maintenance of good health, fish require water of a certain minimum quality. The specific minimum requirements vary from species to species, but there are some general characteristics that apply to all of the species. In freshwater environments the measurement of salinity is not necessary, but careful monitoring of temperature is important because metabolic rate of the fish is so dependent upon the temperature of water within the culture system.

Many water quality variables need to be determined only once or at intervals of several weeks or months. Examples are hardness and alkalinity when their initially measured levels are sufficiently high to be of no special concern. Fish culturists often do not routinely measure pH or nutrient levels in their water, but those measurements may be necessary in instances where those variables can be expected to fluctuate greatly. For example, pH should be monitored in closed recirculating water systems, and nutrient levels should be monitored when a specific fertility level is being maintained. Routine measurements generally include temperature, dissolved oxygen, and in some systems, ammonia.

Physical Conditions of Water

Temperature

Culture system water temperature should be routinely monitored since management decisions with respect to such activities as spawning and amount of feed that should be provided are often made primarily on the basis of that variable.

Water temperature generally depends upon climate, sunlight and depth. All organisms, including fish, posses well defined limits of temperature tolerance with the optimum lying some where in between. The temperature in fish ponds is generally minimum during the early hours of morning and reaches the maximum value in the afternoon showing diurnal fluctuations. The seasonal fluctuations of water temperature influences greatly on the digestive variation of fish. In winter months, the temperature of water compared with that of summer months are greatly different and variable.

Compared to the yields of fish in ponds in temperate zones, the natural water in tropical areas generally show a higher production due to more heat budget in the pond system. Apart from these, temperature plays very important role in physiological processes for breeding mechanisms in fish both under natural and artificial conditions. All biological activities like digestive variations, reproduction, movement and distribution are greatly influenced by water temperature. The chemical changes in both soil and water are greatly influenced by temperature. Decrease in DO_2 is directly related to increase in temperature.

Fish display great variability in their tolerance to temperature. Indian major carps usually tolerate wide range of temperature and are called eurythermal. Fishes of temperate region can survive under ice in winter but those of tropics cannot stand such low temperature. Tilapia cannot stand temperature below 8.9°C, grass carp can with stand temperature up to 40°C. The silver carp shows sign of distress at 40°C. Indian major carps can grow well in the temperature range of 26–32°C.

Temperature can be measured with a glass thermometer containing mercury or alcohol, an in-line temperature gauge, or with a thermistor used in conjunction with a temperature meter, which may also automatically record the readings. Computerized systems have been developed which will automatically adjust the flow rates of warm and cold water through mixing valves and maintain the desired temperature within the system.

Depth

Depth of a pond has an important bearing on the physical and chemical qualities of water. Depth determines the temperature, the circulation pattern of water and the extent of photosynthetic activity. In shallow ponds sunlight penetrate up to the bottom and increase the productivity very shallow ponds get overheated during summer. A depth of about 1–2m is ideal.

Turbidity

Turbidity may be due to suspended inorganic substances such as silt and clay or due to planktonic organisms. Pond with clay bottom are likely to have high turbidity. Turbidity due to silt or mud is harmful to fish and fish food organism. Turbidity restricts the penetration of light and reduces the photosynthetic activity. Fishes dwelling in turbid waters have reduced eyes.

Most undesirable type of turbidity is that resulting from suspended clay or silt particles. Turbidity due to plankton is somewhat desirable, if it does not lead to bloom. Turbidity caused by suspended silt and clay particles can be controlled by application of hydrated lime. Boyd(1979) tested the effectiveness of four coagulants: filter alum, hydrated lime, ferric sulphate and agricultural gypsum. However, most effective chemical to remove high turbidity is filter alum, $Al_2(SO_4)_3$, 14 H_2O. Application of filter alum, at the rate of 25–30 mg/l will remove turbidity from most waters. But alum application reduces pH and alkalinity for which hydrated lime should be applied to the pond before or simultaneously with alum.

Light

Light penetration depends first and foremost upon the available intensity of the incident light. Availability of light energy to a fish farm greatly influences its productivity as primary production is a photo chemical process energized by light. Shade provided by the surrounding vegetation affects the light of the pond. Prevailing turbidity also affect the light penetration.

Chemical Conditions of Water

Alkalinity

Alkalinity is defined as the total amount of acid required to titrate the bases in a water sample. Total alkalinity of a water sample includes carbonate, bicarbonate, hydroxide, silicate, phosphate, ammonia, and various organic compounds; the majority of the bases are attributable to carbonate and bicarbonate, so an alternative definition is that alkalinity is the total bicarbonate and carbonate in a water sample. It is also a measure of the buffering capacity of the water. Measurement of alkalinity is by two-step titration. The first step takes the water sample to its phenophthalein endpoint and measures carbonate alkalinity (reported as milligrams per liter of calcium carbonate ($CaCO_3$)). The amount of titrant used in the second step (indicated with the indicator methyl orange), when added to the first, provides a measurement of total alkalinity (also reported as milligrams per liter of $CaCO_3$). The difference between total and carbonate alkalinity is a measure of the bicarbonate alkalinity.

Most water sources remain constant or nearly constant with respect to alkalinity over long time periods. Occasionally, culture systems are developed with sources of water that are of extremely low alkalinity. When total alkalinity is less than about 20mg/l, it may be necessary to add a source of carbonate to the ponds. $CaCO_3$ can be added to ponds to increase alkalinity (and hardness).

Pond pH is affected by photosynthesis, which removes carbon dioxide (CO_2) from the water, and respiration, which adds carbon dioxide. When CO_2 is dissolved in water, it forms a weak acid (carbonic acid) that tends to drive the pH down. When CO_2 is removed by plants during photosynthesis, there is less acidity and the pH tends to rise.

As long as there is a supply of carbonate and bicarbonate available in the water, hydrogen ions will be released or removed from solution and the pH will not change. This process is known as buffering and can be extremely important. If the pH of the water is not maintained within the ranges of approximately 5.0 to 9.0 some of the species of fish may be killed (Randall 1991). The addition of calcium carbonate to a pond as well as somewhere within closed recirculating water systems provides a source of carbonate, and then bicarbonate as the calcium carbonate goes into solution: The carbonate–bicarbonate buffer system can normally be depended upon to maintain the pH in fish culture ponds within the desirable pH range of approximately 6.0 to 9.0.

Hardness

It is defined as the total soluable calcium and magnesium salts present in the water medium, expressed as its equivalent $CaCO_3$. Hardness does not normally vary temporally, but in instances where the water is extremely soft (contains less than 20 mg/l total hardness), it may be necessary to add gypsum or crushed limestone to provide the cations needed for maintenance of the buffer system. Limestone (calcium carbonate) will increase both hardness and alkalinity, since upon dissociation it provides both divalent calcium and carbonate ion.

In extremely hard water (hardness of a few hundred milligrams per liter), calcium carbonate may precipitate on the surfaces of materials that come in contact with the water. This can cause the eventual constriction of water pipes and make glass aquaria opaque, sometimes after only a few days. Glass can be cleaned with weak hydrochloric acid, but the problem will recur. Pond water having a hardness of 15 ppm or above are satisfactory for the better growth of carps but hardness below 5 ppm leads to slow growth and distress.

Nitrite

Fishes suffering from nitrite toxicity have frequently been observed in recirculating systems in which the bacterial flora has not been adequately established on the biofilter medium prior to fish stocking. Fishes suffering from nitrite toxicity exhibit chocolate–colored blood, hence the common name for nitrite toxicity syndrome, brown blood disease. When nitrite levels become sufficiently high, haemoglobin reacts with nitrite to form methemoglobin which is unable to combine with oxygen. Thus, the tissues are deprived of oxygen and the affected fish are eventually asphyxiated. Fish suffering from methemoglobinemia may begin to swim erratically just prior to dying.

Toxicity at nitrite levels from below 0.2 mg/l to nearly 200 mg/l, depending upon the species being studied. Nitrite toxicity is pH and chloride dependent (Russo and Thurston 1991). Adding calcium chloride or sodium chloride to the water has been an effective means of reducing nitrite toxicity in aquaculture ponds (Tomasso *et al.*, 1979, 1980; Huey *et al.*, 1980; Schwedler and Tucker 1983). Levels of either chemical applied at 60 mg/l have been efficacious.

Dissolved Oxygen

Among the chemical substances in natural waters, O_2 is probably one of primary importance both as a regulator of metabolic processes of plant and animal community and as an indicator of water condition. The pond water receives oxygen mainly through (1) interaction of atmospheric air on the surface water of pond (2) by photosynthesis. The oxygen liberated during photosynthesis is influenced by temperature, light, concentration of dissolved salts, turbulence of water including the abundance of plants.

Transfer of oxygen from air to water and vice versa occurs with the unsaturation and saturation of this element in water respectively.

For average or good production, ponds should have DO_2 concentration above 5.5 ppm. It may be mentioned that very high concentration of DO_2 leading to a state of supersaturation sometimes lethal to fish fries during rearing of spawn in nursery ponds. In the pond systems, DO levels are typically lowest near dawn as a result of continuous respiratory demand by both plants and animals during the night and absence of photosynthetic oxygen production. As photosynthesis begins to introduce oxygen into the water after dawn, the DO level will begin to rise. The DO will often increase throughout the daylight hours and then begin to decline towards dusk.

Some of the factors which can lead to a net reduction in DO are; a period of overcast weather, increased turbidity, and the presence of herbicides in the water. All have a negative effect on phytoplankton productivity, thereby reducing oxygen production. Net increases in DO can occur when phytoplankton productivity is increasing, when there is wind mixing, or when the culturist takes some action to enhance DO.

The fish farmer must routinely check each pond for signs of oxygen depletion in the morning at about dawn, particularly during the peak growing season, when the water is warm, fish densities are high, and the respiratory demand on oxygen is great. Many farmers now monitor DO throughout the night, either utilizing personnel to observe the ponds and collect DO measurements as appropriate, or by employing automated systems to monitor

DO measurements as appropriate, or by employing automated systems to monitor DO concentration.

Fish that are experiencing hypoxia will often appear to be gasping at the surface of the water. By measuring the DO level in the water, action to prevent an oxygen depletion can be taken before the fish begin to exhibit that behaviour, which indicated that they are already severely stressed.

DO can be measured chemically utilizing a method known as the Winkler titration, which is inexpensive. Highly reliable and suitable accurate DO meters are currently available and widely employed by fish farmers. They are normally battery operated and often have associated with them a thermistor for temperature measurement in conjunction with an oxygen electrode.

Emergency aeration procedures to overcome oxygen depletions include the use of paddlewheel aerators or some other type of aeration device, running oxygen-rich new water into the pond, recirculating the pond water through a pump which sprays the water back in forcefully, injecting compressed air or air provided by a regenerative blower, employing agitators of various kinds, or injecting compressed or liquid oxygen.

Dissolved oxygen in pond water should not be less than 3.0 ppm and for normal growth it should be more than 5 ppm. The dynamics of DO_2 is strongly influenced by phytoplankton density. To maintain a desired level of O_2, the organic rich sediment of pond should be removed and also limed. Proper regulation of manuring and fertilization application resulting in required phytoplankton cells per litre of water without causing bloom is necessary. During oxygen depletion, the pond water should be agitated smoothly by splashing.

pH

The pH of a water is defined as the logarithm of the reciprocal of hydrogen ion concentration. It may be expressed mathematically.

$$pH = Log [1/(H^+)]$$

The pH values vary between 0 and 14.

☆ Equal to 7 - Neutral

☆ > 7 - Alkaline

☆ < 7 - Acidic

The pH of pond water shows a diurnal change, it is being alkaline in mid afternoon and acidic just before day break. The best water for fish cultivation is that which is neutral or slightly alkaline between 7 and 8.0. The limit above or below which pH has a harmful effect is given as 4.8 and 10.8. The acidic water reduces the appetite and growth of fishes. The toxicity of H_2S, CH_4 and other heavy metals to fish is increased in increasing pH. Fish gets prone to attack of parasites and diseases in acid waters.

Application of lime raises the pH of soil as well as that of water. Liming establishes a strong buffer system and stimulates release of nutrients through microbial decomposition. The amount of lime to be applied depends upon the pH, texture and organic matter content of the soil. The most common liming materials are agricultural limestone $CaCO_3$ or $CaMg(CO_3)$, hydrated lime, $Ca(OH)_2$, and quick lime CaO. Of these, ground agricultural limestones are extensively used in fish ponds. Neutralising effect of $CaCO_3$ is 100 per cent. Considering this the value for $CaMg(CO_3)_2$ is 109, $Ca(OH)_2$ is 136 and for CaO is 179. Liming can be applied to the dried pond soil or broadcasted over water surface either in single installment or equal split doses. In certain fish ponds the pH value found to be 11.0 or 12.0 indicating highly alkaline water. In this case the pH of pond water may be reduced using agricultural gypsum ($CaSO_4$, $2H_2O$). The rate of application of gypsum depends upon the total alkalinity and calcium hardness of water.

Ammonia

Ammonia exists in two states, ionized ammonia (NH_4^+) and unionized ammonia (NH_3). The sum of the two is called total ammonia or simply ammonia. In most environments NH^+_4–N predominates. However, it depends on pH, temperature and salinity. An increase of one pH unit will generally cause a tenfold increase in the proportion of NH_3–N solution systems with high pH are more prone to ammonia toxicity. In seawater, the mole fraction of NH_3–N is about 40 per cent less than that of freshwater at the same pH and temperature.

Unionized ammonia, NH_3–N is very toxic to fish and invertebrates, even in very small quantities. An ambient ammonia concentrations increase, the ability of fish to excrete ammonia decreases, and the level of ammonia in the blood and tissue increases. This results in elevated blood pH and adverse effects on enzyme reactions and membrane stability. The effects include gill damage, reduction in the O_2 carrying capacity of the blood, increased O_2 demand by tissues and histological damage to the red blood cells.

Ion exchange is an efficient and reliable means of ammonia removal, but it is much more expensive than biological filtration. Clinoptilolite is one of the zeolites used in water treatment, is an effective natural ion exchange material for the removal of ammonia from hatchery water.

There is no practical method to reduce concentration of ionized or unionized ammonia in large, commercial fish ponds. In smaller systems dilution by water exchanges and lowering the pH can reduce. The only method to reduce TAN with some degree of performance is to reduce stocking and feeding rates.

Dissolved Solids

In natural aquatic systems, water contains both inorganic and organic dissolved solids. Inorganic solids in solution contains anions of CO_3, Cl, SO_4, PO_4, NO_3 in combination with Ca, Na, K, Mg, Fe etc. Organic compounds are in the organic state of P, N, Sugar, acids and vitamins. Both organic and inorganic dissolved solids in pond water varies qualitatively and quantitatively with the season. The edaphic relationship that contributes to the productivity of the water is dependant to the total concentration of dissolved solids in a water.

Chapter 10
Feed and Feeding Management

The 'manufactured' foods can be broadly classified as dried, freeze-dried and frozen diets. Dried foods such as flakes, pellets, sticks and tablets, are widely available and are convenient to use. Many types of dried foods, especially flaked foods, feature special 'growth', 'colour' and 'vegetable' diets. Most dried foods have a limited shelf life and generally they should be used within a few months of purchase. Although they can still be used after this period, the vitamin levels decrease with time.

Nutrients

Proteins (amino acids), lipids (fats, oils, fatty acids), carbohydrates (sugars, starch), vitamins and minerals are important nutrients required for the better growth and health of fishes.

Proteins

These are complex organic compounds composed of a number of amino acids linked together through peptide ponds. The growth of fish in terms of muscle formation depends on the protein intake. The protein requirement for optimum growth is between 35 and 40 per cent for carps.

Fry of carp needs feed with 50 per cent protein. Fish meal, soybean meal, skim milk powder, legumes, and wheat gluten are excellent sources of protein. Additionally, the building blocks of proteins (free amino acids) such as lysine and methionine are commercially available to supplement the diet.

Lipids

The fats and oil are high energy yielding substances. Fish feed for profitable fish culture should have 4-18 per cent of fat. Oils from marine fish, such as sardines, and vegetable oils from canola, sunflower and linseed, are common sources of lipids in fish feeds.

Carbohydrates

10-15 per cent of Carbohydrates in fish feed enhances the growth of fish through their protein sparing action. Cooked carbohydrates, from flours of corn, wheat or other "breakfast" cereals, are relatively inexpensive sources of energy that may spare protein (which is more expensive) from being used as an energy source.

Vitamins and Minerals

The variety and amount of vitamins and minerals are so complex that they are usually prepared synthetically and are available commercially as a balanced and pre-measured mixture known as a vitamin or mineral premix. This premix is added to the diet in generous amount to ensure that adequate levels of vitamins and minerals are supplied to meet dietary requirements.

Binding Agents

Another important ingredient in fish diets is a binding agent to provide stability to the pellet and reduce leaching of nutrients into the water. Beef heart has traditionally been used both as a source of protein and as an effective binder in farm-made feeds. Carbohydrates (starch, cellulose, pectin) and various other polysaccharides, such as extracts or derivatives from animals (gelatin), plants (locust bean), and seaweeds (agar, caragenin, and other alginates) are also popular binding agents.

Preservatives

Preservatives, such as antimicrobials and antioxidants, are often added to extend the shelf-life of fish diets and reduce the rancidity of the fats. Vitamin E is an effective, but expensive, antioxidant that can be used in laboratory prepared formulations. Commonly available commercial antioxidants are butylated hydroxyanisole (BHA), or butylated hydroxytoluene (BHT), and ethoxyquin. BHA and BHT are added at 0.005 per cent of dry weight of the diet or not more than 0.02 per cent of the fat content in the diet, while ethoxyquin is added at 150 mg/kg of the diet. Sodium and potassium salts of propionic, benzoic or sorbic acids, are commonly available antimicrobials added at less than 0.1 per cent in the manufacture of fish feeds.

Attractants

Other common additives incorporated in the fish feeds are chemo attractants and flavorings, such as fish hydrolysates and condensed fish solubles (typically added at 5 per cent of the diet). The amino acids glycine and alanine, and the chemical betaine are also known to stimulate strong feeding behaviour in fish. Basically, attractants enhance feed palatability and its intake.

Table 11: Nutritional Requirements of Small and Adult Fishes:

	Protein	*Fat*	*Carbohydrate*
Small fishes	40-45 per cent	4-6 per cent	40 per cent
Adult fishes	30-35 per cent	6-8 per cent	50 per cent

Artificial feed is classified into three major categories according to the preparatory methods.

1. Dry Feed

The feed is in dried form with 8-10 per cent moisture. It is further classified into 5 sub groups.

- ☆ Pellets
- ☆ Flakes–are flat in structure.
- ☆ Freeze dried feed–They are in frozen form
- ☆ Tablet form
- ☆ Granular form–they are very small and round shape. They are similar to grains.

2. Moist Feed

It is prepared daily and fed to fishes–the moisture content is 35 per cent.

3. Paste Feed

This is mainly prepared for young ones of all. The feed ingredients are made in to paste and fed to fishes squeezing through mesh.

Feed Formulation

Since protein plays a significant role on the growth of fishes, protein level is given importance while formulating the feed. To balance and formulate the required protein diet Pearson square method is followed.

Using Two Ingredients

Using rice bran (protein 9 per cent) and ground nut oil cake (protein 45 per cent) a fish diet with 27 per cent protein can be prepared as follows.

A square is constructed and the two ingredients are put on the two left corners along with the protein content of each. The desired protein level of the feed is placed in the middle of the square. Next, the protein level of the feed is subtracted from that of the ingredients and the answer is placed in the opposite corner.

Rice Bran

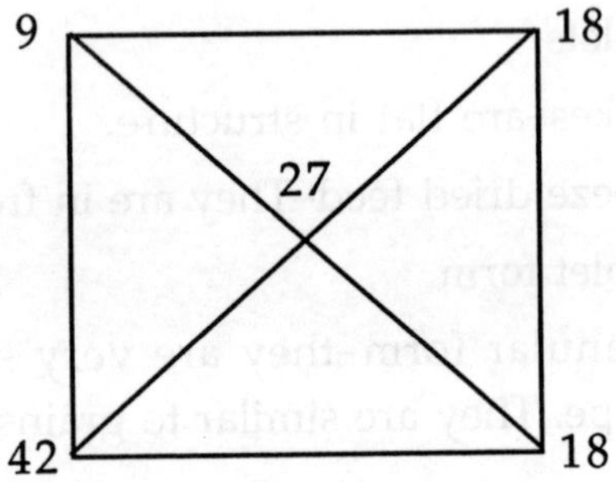

Groundnut Oil Cake

Now the figures on the right side corner of the square are added *i.e.* 18 + 18 = 36

% of rice bran to be added in the feed = $\frac{18}{36} \times 100 = 50\%$

% of Groundnut oil cake = $\frac{18}{36} \times 100 = 50\%$

To make 1 kg feed 500g rice bran and 500g-groundnut oil cake should be added.

Using More than Two Ingredients

Using prawn waste (Protein 35 per cent), fish meal (protein 60 per cent), wheat bran (Protein 15 per cent) and tapioca (Protein 2 per cent), a fish diet with 32 per cent protein can be prepared as follows.

The ingredients with less than 20 per cent protein are grouped based on their protein content first.

Wheat bran	=	15 per cent
Tapioca	=	2 per cent
		17 per cent
Average	=	17/2 = 8.5 per cent

The ingredients with more than 20 per cent protein are based on their protein content are also grouped subsequently.

Fish meal	=	60 per cent
Prawn waste	=	35 per cent
		95 per cent
Average	=	47.5 per cent

Now the determined average protein contents are put on the two left corners of the square and rest is same as that of previous method.

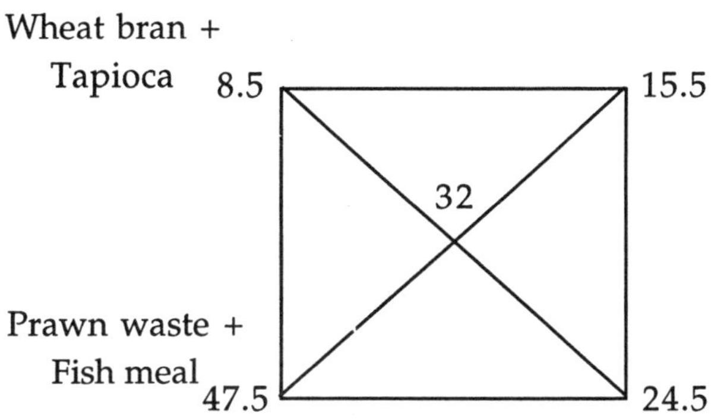

Now, the figures on the right hand side corners of the square are added as follows.

$$15.5 + 24.5 = 40.0$$

Wheat bran + tapioca = $\frac{15.5}{40} \times 100 = 37.75$

Prawn waste + fish meal = $\frac{24.5}{40} \times 100 = 61.25$

Wheat bran = $\frac{38.75}{2} = 19.375\%$

Tapioca = $\frac{38.75}{2} = 19.375\%$

Prawn waste = $\frac{61.25}{2} = 30.625\%$

Fish meal = $\frac{61.25}{2} = 30.625\%$

The composition of the feed is

Prawn waste	=	30.625%
Fish meal	=	30.625%
Wheat bran	=	19.375%
Tapioca	=	19.375%
Total	=	100%

After deciding the composition of feed mixture, ingredients are ground in a mixie individually and then they are mixed together. Cooking is done with a water addition at 30-35 per cent level. The dough is thus prepared.

Flow Chart for the Preparation of Feed

Feed ingredients

⬇

Sieved through 300 micron mesh

⬇

Mixing

⬇

Addition of water 30–35 per cent

⬇

Dough

⬇

Cooking for 15 min in pressure cooker

⬇

Cooled

⬇

Oil + Vitamin + Antioxidant + growth promotion + Antibiotic sprayed over the feed.

⬇

Remixing

⬇

Pressed through hand pelletiser (2mm)

⬇

Drying

Feeding Management

- ☆ A quantity of food given at a time would have been eaten entirely before the next distribution.
- ☆ Distribution should be preferable in the morning.
- ☆ The food should not be spreaded out over the whole surface of the pond but it is placed at different marked spots.

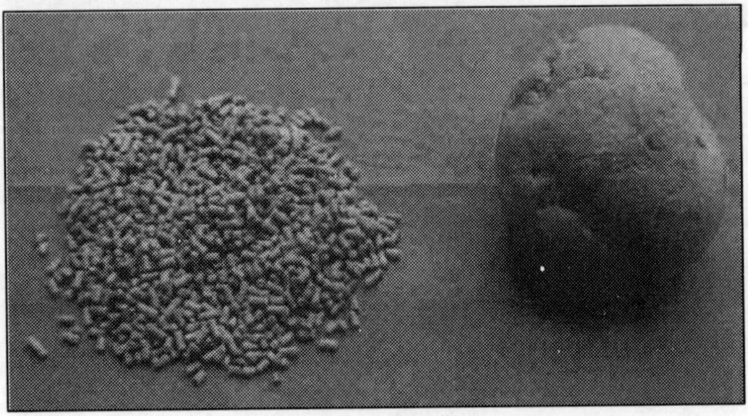

Figure 8: Artificial Feed

- ☆ The spots chosen for distribution should be clean and should have 0.6-1 m depth. There should be no vegetation around the spot. The bottom should be hard and not muddy.
- ☆ If varied artificial feeding is practiced, then the variation should be gradual not drastically or suddenly.
- ☆ The amount of food distributed at one time depends on the temperature also.
- ☆ In the case of epizootics, feeding ceases. The amount of food given at this time should be considerably reduced.

Chapter 11
Culture of Carp

Our aquaculture is mainly carp-based where in catla, rohu and mrigal are grown together under polyculture system or along with the 3 exotic carps, *viz.* silver carp, grass carp and common carp, as composite carp culture systems. These 6 species are selected because of their compatibility for habitat preference and food to utilize all the ecological niches of the culture system. While catla and silver carp are basically surface feeders showing preference for zoo- and phytoplankton, respectively, mrigal and common carp are omnivorous bottom feeders. Rohu is a column feeder and grass carp shows high preference for submerged vegetation. Feeding at the base of the food chain, phyto- and zooplankton, detritus and aquatic weeds, these carps utilize natural productivity with greater output. However, among 6 species, the Indian major carps are slow growing as compared to their exotic counterparts. Thus, polyculture of 3 Indian carps usually results in lower production level than six-species composite culture system or the system with 3 exotic carps. These carps are also culture under mixed culture system incorporating freshwater prawn (*Macrobrachium rosenbergii, M. malcolmsonii*)

Carp culture is undertaken mostly in earthen ponds of varying dimensions. Several combinations of cultural practices have been evolved in the country to suit fish species, water resources, availability of fertilizer and feed resources. The standardized packages of practice for carp polyculture includes:

1. Pond preparation
2. Liming
3. Fertilization
4. Stocking
5. Feeding and Feed management
6. Water-quality management and
7. Health management.

Grow-out culture of carps is carried out in earthen ponds ranging from 0.05 to 5.0 ha and 1-3 m in depth in different regions of the country; and ponds of 0.5-1.0 ha size with water depth of 1-2 m is considered ideal. While small and shallow stagnant ponds have several inherent problems, which adversely affect fish growth, the large and deep ponds have their own problems of management. Essentially, management practices in carp polyculture involves environmental and biological manipulation for obtaining higher levels of fish production, which can be broadly classified as pre-stocking, stocking and post-stocking operations.

Pre-stocking Management

Pre-stocking management includes the following practices:

- ☆ Eradication of aquatic weeds
- ☆ Eradication of predatory fishes
- ☆ Eradication of weed fishes
- ☆ Control of aquatic insects
- ☆ Liming and
- ☆ Fertilization

Fertilization Practices

Pond fertilization aims at enhancement of autotrophic and heterotrophic production of ecosystem, which is stimulated by organic and inorganic fertilizers. In carp culture, the usual practice involves application of both organic manures and nitrogenous and/or phosphatic fertilizers. Freshwater ponds of the country usually possess adequate amount of potassium required for natural productivity, thus potassic fertilizers are usually not supplied to culture ponds. Cattle-dung or poultry droppings are most commonly used organic manures, often used in combination with urea and super-phosphate as inorganic nitrogen and phosphorus sources. The conventional dosage followed in carp culture practice in India usually ranges from 10 to 20 tonnes of cattle dung/ha/year of 4-8 tonnes of poultry manure/ha/year alone or in combination with urea @ 100 kg N/ha/year and superphosphate @ 50 kg P/ha/year. Of the total amount of organic manures, while 25-30 per cent is generally applied as basal dose a fortnight before stocking, the remaining is applied in equal installments at fortnightly intervals. Beside cattle-dung and poultry droppings, other organic manures with high nitrogen and phosphorus contents such as pig manures, duck droppings and domestic sewage are also used depending on the availability.

Figure 9: Composite Fish Culture

Attention has been shifted lately towards the use of biofertilizers and bioprocessed organic materials for ensuring sustainability of aquaculture practices and to avoid possible environmental materials have been tested for their efficacy. *Azolla*, a nitrogen-fixing aquatic fern, with high protein content (15-17 per cent), has been standardized as an ideal nitrogenous biofertilizer for aquaculture at 40 tonnes/ha/year; providing almost full complement of nutrients required for intensive carp culture (100 kg nitrogen, 25 kg phosphorus, 90 kg potassium and 1,500 kg organic matter). The detritus resulting from decomposition of *Azolla* applied in ponds serves as a trophic component of carps and prawns. The bioprocessed organic manure, biogas slurry, has been standardized as a manure in carp culture, @ 30-45 tonnes/ha/year, with distinct advantages in terms of lower oxygen consumption and faster nutrient liberation rates. The use of *Azolla* and biogas slurry has been found economical compared to traditional fertilization practices.

Stocking

The stocking density of carps for a culture pond is predetermined based on the targeted production, pond productivity and carrying capacity, species to be cultured, their feed-conversion efficiencies, size at stocking, growing period and level of management. Though fingerlings of more than 100 mm constitute best stocking material for grow-out culture, constraints in their availability often force farmers to resort to stoking of smaller fry. In intensive culture, a size of 50-100 g is preferred for stocking to realize higher survival and better growth. Generally a density of 5,000-10,000 fingerlings/ha is kept as a standard stocking rate in carp polyculture for a production target of 3-5 tonnes/ha/year. In seasonal ponds or where water level becomes limiting during summers, it is reduced to 2,000-3,000 fingerlings/ha to obtain higher growth. Although major carps are expected to reach an average of 0.8-1.0 kg in the first year, the growth rate is invariably reduced at higher stocking densities. With provision of water exchange and aeration, higher targeted fish production levels of 10-15 tonnes/ha/year are achieved by resorting of stocking ponds at a density of 15,000-25,000/ha.

Selection of compatible species and stocking them at appropriate ratio suiting to culture environment is another important criteria in carp culture. Manipulation of species ratio is done for minimizing the interspecific and intraspecific competition for available food at various trophic levels in a pond. More than one species occupying different niches could be utilized in a pond for exploiting available natural food. Several combinations of Indian major carps only and in combination with exotic carps have been experimented under different stocking densities keeping growth as an index of performance. A combination of catla, silver carp, rohu, grass carp, mrigal and common carp fulfils the criteria of species, and has proven an ideal combination for freshwater carp culture in India. A proportion of 30-40 per cent surface feeders (silver carp and catla), 30-35 per cent mid-water feeders (rohu) and 30-40 per cent bottom feeders (common carp and mrigal) is commonly adopted depending on the productivity of the pond. When there is a possibility of providing aquatic vegetation as food for grass carp on a regular basis, the same may be included at 5-10 per cent of the total. While ponds with higher organic matter can be stocked with higher proportion of bottom feeders, surface feeder may be kept more in newly excavated ponds. Similarly, rohu may be stocked at higher ratio in ponds with higher water depth.

Supplementary Feeding

The natural productivity of pond, irrespective of the level of its augmentation through fertilization, is not capable to sustain a higher level of fish biomass in semi-intensive grow-out. Due to the limitation in availability of natural fish food in pond at higher stocking density, the energy requirement for somatic growth can be met only through provision of supplementary feed. Over years, nutritional requirement of carps has been studied. Various ingredients of plant and animal origin, mixed at various combinations, have been evaluated in culture systems for production performance of the carp species. Among those, mixture of groundnut/mustard oilcake and rice-bran at 1: 1 has been commonly used in carp culture.

Several balanced feed formulations have been developed making up deficient essential amino acids, fatty acids, and incorporating vitamins and minerals. All these ingredients are blended together at required proportion for preparation of carp feed. The commercial production of supplementary feed is usually in pellets of different diameter, to provide higher water stability, consumption and utilization by the fish. Under feeding depresses fish growth while over feeding results in wastage of food, leading to deterioration of water quality. As feed cost constitutes more than 60 per cent of the recurring expenditure in carp culture feed management, is an important aspect for ensuring optimum growth of carps and limiting production cost, besides maintenance of pond aesthetic condition. The recommended practice is to provide feed at 3-5 per cent of body weight of stocking material initially and subsequently at sliding scale from 3 to 1 per cent.

The daily ration is adjusted according to the biomass estimated from monthly sampling and it is provided in two splits during morning and evening in feeding trays or perforated gunny bags suspended at regular intervals in pond. Grass carps are provided with aquatic vegetation like duckweeds, *viz. Lemna, Spirodela* and *Wolffia; Hydrilla, Najas* and *Ceratophyllum* at periodical intervals. To overcome shortage in supply of desired vegetation with heavy demand consequent on low conversion rate of aquatic vegetation to fish flesh, marginal vegetation, land grasses and other fodder, banana leaves and vegetable refuse can also be used.

Management of Soil-and-Water Quality

Management of soil-and-water quality in carp culture ponds is an integral part of the culture operation for a successful crop. A number of environmental problems are sometimes encountered in grow-out ponds. Some of these are inherently associated with the site characteristics like porosity, acidity and high organic-matter content of bottom soils, and some are encountered during cultural operations. Drying of pond bottom to crack between crops helps in aeration, which enhances microbial decomposition of soil organic matter. The porosity of the pond bottom is corrected through bentonite, lining with plastic sheet at 0.3-0.5 m depth and heavy doses of organic manures (cattle-dung at 10,000-15,000 kg/ha/year).

The water-quality parameters required for optimum growth of carps are pH 7.5-8.3, temperature 27-32°C, dissolved oxygen more than 4 mg/l, total alkalinity of 80-200 mg $CaCO_3$/l, Secchi disk visibility of 25-30 cm, total inorganic nitrogen 0.5-1.0 mg/l and phosphorus 0.2-0.3 mg/l. Variations in these water parameters occur in culture pond and need periodic correction/management measures. The low water pH is corrected through intermittent application of lime materials. Lime helps in improving alkalinity, hardness, controlling turbidity and reducing H_2S build-up. Dolomite is particularly useful when augmentation of phytoplankton growth is required. Agricultural gypsum ($CaSO_4$) is applied to correct alkaline pH. It is also applied to increase total hardness without affecting alkalinity. Aeration, a proven method for improving pond dissolved oxygen availability, also helps in mineralization process reducing organic load. Water exchange helps to a great extent in reducing metabolic load. Addition of water into pond, particularly during winter, helps in improving temperature regime and prevents temperature stratification. Turbidity with suspended soil particles can be controlled by cattle manure (500-1,000 kg/ha), gypsum (250-500 kg/ha) or alum (25-50 kg/ha). The ammonia load in pond can be reduced through encouraging healthy growth of phytoplankton or aeration. The H_2S build up in pond can be subsided through frequent water exchange.

Health Management

A balanced relationship among the host, the pathogen and the environment is a prerequisite for optimum growth and health of any organism. Any imbalance in this relationship leads to diseased condition of the host, *i.e.* animal. Fish is a cold-blooded animal and more prone to environmental deterioration. In the event of a disease outbreak, it affects fishes in groups or the whole population. Further, treating a diseased fish is relatively difficult compared to terrestrial animals, as individual treatment is not possible in former, thereby requiring mass treatment of population or environment. Therefore, prevention is always a preferred method in aquaculture to control disease outbreak than curing disease. In recent years, due to increase in intensity of culture, disease management and health care has become an important aspect of management for preventing sudden outbreak of epizootics that occur due to environmental deterioration, improper feeding and, overcrowding.

Underfeeding leads to malnutrition, resulting in growth retardation and low disease resistance. Many of the fishes may carry small number of pathogens like bacteria, virus, fungi and parasites, either as chronic low-grade infections or serving as carriers. Such pathogens are also present in deteriorated pond-water and sediment and their population increases to a very high level when pond is heavily stocked. Weak disease resistance of the fish makes them susceptible to attack from these pathogens.

The best way to avoid disease outbreak in pond during culture is through taking preventive measures which are ensured by proper management of the soil-and-water quality, following proper feeding schedule, use of balanced feed, periodic sampling for health check, and minimizing outside influence on the pond.

Harvesting and Marketing

Generally carps are harvested after a grow-out period of one year during which it reaches marketable size of 0.8-1.0 kg. However, these carps are even marketed in smaller sizes of over 300 g, except that of silver carp; the marketable size for which is over 1 kg. In multi-harvesting system, the fishes attaining market size are periodically harvested from the pond, releasing back smaller ones for further growth. While intermittent harvestings are done with dragnet of suitable mesh size, final harvesting is usually done by complete draining of pond.

Carps produced from culture ponds are mostly sold in local market, either in live or dead condition. The Indian major carps are also transported to adjacent deficit areas as well as to distant places, even 2,000-3,000 km away from the production site, in insulated vans with ice. Fresh fish fetches about one-and-half-times higher market price than

Figure 10: Harvesing of Carps

iced ones. When sold in live condition, the carps command still higher sale value of over two times compared to iced ones. The price in the domestic market is influenced by demand and supply.

Chapter 12

Culture of Tilapia

Tilapias belonging to family Cichlidae, are natives of Africa. They have been introduced into a large number of tropical and sub-tropical countries during the period of last four to five decades, either accidentally or deliberately. Jhingran and Gopalakrishnan (1974) listed 22 species that have been used in experimental or production-scale fish culture. The substrate spawners, which make nests on the bottom of water bodies and spawn in them retained the name *Tilapia* and the mouth-brooders, which incubate the fertilized eggs in the mouth of the female or male parent, came under a new genus *Sarotherodon* (meaning 'brush toothed') (Trewavas, 1982). Later, a new genus Oreochromis was constructed to accommodate species which spawn in nests on the bottom of water bodies, but brood the eggs in the mother's mouth. For commercial aquaculture, the most important species of tilapia are: *T. rendalli, T. zillii, T. mossambica, T. melanotheron.*

Tilapias are euryhaline fishes and grow well both in fresh and brackish waters. Species like *T. mossambica and T. zillii* can even grow in hypersaline waters of 42 ppt salinity. Tilapias are mostly herbivorous or omnivorous. They have remarkably hardy and disease-resistant and tolerate wide ranges of environmental conditions. Tilapias are probably the most widely cultured fish species in the tropics. In Latin American countries, the more popular species are *T. niloticus, T. rendalli, T. aurea* and *T. mossambicus.* Indigenous species are more popular in different African countries and several species besides the above mentioned have given very promising results in experimental farms, such as *T. macrochit, T. niger* and *T. andersonii* in Zambia and Uganda, *T. melatheron* and *T. guineensis* in Ivory coast, *T. sheranus* in Malawi, etc.

Being herbivorous or omnivorous, it was comparatively easy to feed the species of this group. They were found to be hardy and could be reared in fresh, brackish and even sea water. Despite some of the advantage of tilapia as candidate species for small or large-scale culture, it was soon realized that the culture technique for producing marketable fish was not as easy as originally believed. Their early maturation and frequent breeding, especially in tropical climates, affected growth rates to such an extent those special measures of stock management and feeding were found necessary to produce fish suitable for human consumption. The most common and widely practiced system of culture of tilapia is in earthen ponds and similar impoundments. As the species can survive in a restricted space, all sizes of ponds have been used, including those measuring less than 100 m^2.

Attempts have been made to develop suitable seed production systems for tilapias. A relatively simple method of producing large numbers of *T. nilotica* fry is by using

Figure 11: Harvest of Tilapia

small earthen ponds (0.05 ha) with about 40 cm water depth. Female fish of about 700 g and males of 200 g are stocked at the rate of 2 fish/m^2 in the male : female sex ratio of 1 : 4-5. Fishes are fed on a high-protein diet for about a month by which time they commence spawning, the brooders are transferred to the second pond where they are fed in the same way as in the first pond. Feeding is continued in the first pond for another month, by the end of which the fry reach a size of about 4 cm and on an average about 5,000 fry are available for harvest. By this time, spawning occurs in the second pond and the brood fish are transferred back to the first pond for further spawning. The production per month by this method is reported to be about 4 to 5 fry/m^2.

In Israel, pond breeding of *T. nilotica* and *T. aurea* is carried out. The ponds are dried prior to spawning in order to eradicate predators. The pond is then filled to a depth of 50-60 cm which is a desirable requirement for the breeding of these species. As the number of eggs per spawning depends on the size of the females, the rate of stocking has to be adjusted in accordance with their size. While a 100 g female *T. nilotica* sheds about 100 eggs, the females of 600 to 1,000 g yield 1,000 to 1,500 eggs. A female *T. aurea* of 1,000 g shed 2,000 eggs at each spawning. The males are usually stocked at the rate of 100 to 250 nos/ha.

In the Philippines for spawning and fry rearing, ponds or cages or hapas are used. The male : female sex ratio of 1 : 3 works all right for spawning. The fish breeds throughout the year. The fry are collected periodically and grown to fingerling size in nursery ponds or cloth tanks. Nurseries are stocked at the rate of 50,000 to 1,00,000 larvae/ha, depending on the size of fingerlings to be raised.

With low stocking densities and intensive fertilization and feeding tilapia is cultured in ponds at the rate of 3,000-5,000 fry/ha. Two to three crops are raised every year. After each harvest, ponds are drained completely.

To control the fry produced by wild spawning in pond culture predatory fishes (2 to 10 per cent of the stock) like *Hemichromis fasciatus, Lates niloticus, Clarias lazera, Micropterus salmoides,* and *Channa striatus* are recommended to be co-stocked. In brackish and saltwater ponds, predatory fishes like *Elops hawaiensis* and *Dicentrarchus* spp. are used.

Often a ban has been imposed on the introduction of tilapia at several places. This is largely because tilapias breed at an enormous rate and virtually sap the water body of all its food material. The result is over-population of tilapia which does not permit other species of fishes to survive nor does it let the tilapia itself grow to an acceptable size.

In order to reduce breeding and over-population, mono-sex culture of males is carried out in a number of tilapia farms. Male tilapia grows at a much faster rate than the female, and under favorable conditions these grows to 500-800 g each. All male culture of tilapia is carried out by adopting sex-reversal technique. For this, the fish are fed with steroid-supplemented diet from the fry stage itself. In about a month or so, the females change to males. Although the technique of producing mono-sex stocks by hybridization or by sex-reversal has been successfully tried at many places, it has not been perfected to the extent necessary to ensure complete absence of female fish.

Mono-Sex Seed Stock and Hybrids

One of the methods in controlling wild spawning of tilapias is mono-sex culture, and since the male tilapia grows faster and attains a larger size, interest has been focused on producing all-male seed stock. Obviously the simplest means is to sort out males from unsorted stocks of fry, commonly known among fish culturists as hand sexing (or manual sexing). The sexes can be distinguished by visual examination of the urino-genital papillae. In the females, the papilla has two orifices whereas the male has only one. Often the female has a smaller genital papilla. No doubt this needs some personal skill and carefulness and can be done reliably only with fingerlings of 20-50 g size. Even at this size, there is likely to be a certain percentage of error in sorting, and even a small number of females in the stock can initiate wild spawning in production ponds.

Another approach to the production of mono-sex stock has been by the use of steroid hormones to achieve sex reversal. It has been possible to reverse the sex of genotypic females by the administration of methyltestosterone or ethynyltestosterone in *T. mossambica*, *T. nilotica* and *T. aurea* (Guerrero, 1982). The degree of success varied between 90 and 100 per cent. The steroids were incorporated in the feed of fry at rates ranging from 10 to 60 mg per kg fish, for durations varying from 18 to 60 days. Feminization of genotypic males was achieved in 90-100 per cent of the males of the above mentioned species, by the administration of estrogens (ethynyloestradiol, oestrone and diethylstilbestrol). The dose consisted of 50-100 mg/kg of ethynyloestradiol (in experiments with *T. aurea* a 100 mg/kg dose of methallubure was included), 200 mg/kg of oestrone and 100 mg/kg diethylstilbestrol and the duration of treatment varied between 19 and 56 days. From comparative experiments, Hanson *et al.* (1983) found that the sex-reversed male populations have a higher growth rate than hybrids and females.

Hopes of using inter-specific hybrids as a means of controlling wild spawning were aroused by the production of all-male progeny by crossing *T. mossambica* females with *T. hornorum* males by Hickling (1960). Besides producing mono-sex male populations, cross-breeding could help in improving catch ability, growth rate, temperature tolerance and body coloration. These could greatly enhance the value of tilapia as candidate species for large scale fish culture.

Since then a number of all male or predominantly male hybrids have been produced.

- ✩ *T. nilotica* × *T. hornorum* (Pruginin and Kanyike, 1965)
- ✩ *T. nilotica* × *T. aurea* (Fishelson, 1962)
- ✩ *T. nilotica* × *T. variabilis* (Pruginin, 1967)
- ✩ *T. spilurus niger* × *T. hornorum* (Pruginin, 1967)
- ✩ *T. vulcani* × *T. hornorum* (Pruginin, 1967)
- ✩ *T. vulcani* × *T. aurea* (Pruginin, 1967)
- ✩ *T. nilotica* × *T. macrochir* (Lessent, 1968)

One hybrid that has received special attention from fish culturists for some time is the so called red tilapia, the colour of which is a blend of pink, yellow and gold. It is appreciated in the market in preference to the normally silvery grey or balck-coloured tilapia. Red tilapia is known to have a faster growth rate and food conversion ratio. It can grow in both fresh- and brackish-water environments. The origin of this hybrid is not yet fully documented. A reddish-orange F-2 progeny with superior qualities was obtained in Taiwan by crossing a mutant reddish-orange female of *T. mossambica* with a normal-coloured grey male *T. nilotica*. In the Philippines, a similar reddish-orange or golden progeny was obtained by cross-breeding a female hybrid of *T. mossambica* × *T. hornorum* with a strain of *T. nilotica*.

All male stocks eliminate many of the problems of tilapia culture, in tropical as well as sub-tropical waters. They can be grown to a larger size of 400-600 g weight to suit consumer acceptance. The longer grow-out period is compensated by the higher prices that large tilapia fetch in the market. Stocking densities of hybrid or all-male populations depend on the level of inputs and management. At even the stocking densities of 3000-5000 per ha, with supplementary feeding of protein-rich leads, individual weight increases of up to 1.5 g per day have been recorded. All-male *T. nilotica* and all-male hybrids of *T. nilotica* × *T. hornorum* of 60-63 g weight, stocked in densities of 10000/ha and cultured for a 6-month period, yielded 2.8 tons/ha of *T. nilotica* and 12 tons/ha of hybrids (Lovshin, 1977). The average weight of the fish was about 240 g. The stocking levels of 13000-31000 per ha of 22-25 g hybrids, annual productions of 5.6-12 tons per ha (of fish weighing up to 495 g) have been recorded (Lovshin, 1982). With stocking densities of 20000/ha and intensive feeding, production of up to 25 tons/ha per annum has been obtained (Hepher and Pruginin, 1982). In such high-density cultures, it is necessary to aerate the ponds, at least during the night.

In Israel, all-male tilapia (*T. nilotica* and *T. aurea*) or hybrid tilapia are commonly cultured in polyculture systems with common carp, mullet and silver carp. Such a combination is believed to contribute to the maximum utilization of feeds and improvement in the chemical regime (dissolved oxygen, ammonia and nitrite levels) in ponds, due to a proper balance of phytoplankton communities and detritus, and also to the better growth rates of some of the species, especially common carp (Sarig, 1983). The proportion of tilapia varies very much between farms, but it is usually higher than mullet and silver carp, and is second only to common carp. Some farms cull a good percentage of carp towards the end of the rearing period and replace them with tilapia. Tilapia then becomes the predominant species. The annual yield in many farms in polyculture can reach 7.5-10.7 tons per year. Tilapia may account for anything between 9 and 53 per cent of the production, the average being around 25 per cent.

At many places, tilapia is cultured in a polyculture system. It has been successfully cultured with fishes such as, carp, grey mullet (*Mugil cephalus*)and African catfish (*Clarias lazera*).

Detritus forms a good percentage of food for the tilapia cultured in ponds and it is an important species in integrated fish-livestock farming systems in several Asian and some African countries.

Cage culture of tilapia has also received considerable attention. Necessary fry or fingerlings for stocking are produced in land-based facilities like ponds, cisterns, and hapas. Two types of cages are used–fixed and floating. Fixed cages are used in shallow eutrophic lakes and the floating cages in deep waters. The stocking density of 25 fingerlings of 3 to 4 cm length/m^2 is common for floating cages. Artificial feeding is generally not practiced except in waters with low productivity. The growth rate is largely based on the primary productivity of the water body and the management practices. The stocking rate in fixed cages ranges from 15 to 50 fingerlings/m^2 and duration of culture ranges from 4 to 12 months.

Pen culture of tilapia is practiced in the Philippines in Laguna de Bay. The same materials which have been used for milkfish pens, are used for the construction of tilapia pens. But for culturing tilapia, smaller pens of 0.5 to 1 ha are preferred. The pens are stocked at the rate of 20-50 fingerlings/m^2. Fishes are fed with rice bran or wheat bran, 2-3 per cent body weight/day. In productive waters, they can grow to a size of 170-250 g in four to five months, even without feeding.

In the Philippines and Thailand, culture of tilapia in rice fields is practiced on an appreciable scale. To obtain marketable fish within a short period of rice cultivation or between two crops of paddy, mono-sex culture of tilapia is adopted.

Chapter 13
Culture of Catfish

Clarias batrachus (Asian Catfish)

Clarias batrachus is the most extensively cultivated species of catfishes in Asia. Its culture is well established in Thailand. In South Asian countries, it forms a major component of live fish culture. Owing to its hardy nature, fast growth and ability to accept artificial feed readily, *C. batrachus* is an ideal fish for aquaculture. It is mostly grown on monoculture basis. The seed of *C. batrachus* can be procured from nature and it can also be produced by breeding the fish under controlled conditions with stimulated natural conditions. The spawning season of *C. batrachus* starts with the onset of the monsoon, the peak period of which is May-September. The rains facilitate migration to the rice fields where these fish normally construct nests in the grassy bed, in shallow water 20-30 cm deep. Females build the nest in the form of round holes measuring 20 cm in diameter and 25 cm in depth. Males take care of the eggs after spawning.

Clay soils with a pH of 6-8 are desirable for spawning ponds, which are normally constructed near the irrigation

Figure 12: *Clarias batrachus* (Asian Catfish)

canals so as to obtain a ready supply of water during breeding. Ponds are usually rectangular in shape and vary in size form 1 to 3 hectares. Along their margin a broodstock holding channel, 2-3 m wide and 1-1.25 m deep, is dug out. In the middle of the pond raised platforms about 2 m wide are constructed in rows with 50 cm wide channels in between them. Grasses are grown on these platforms and holes are made, measuring 20-25 cm in diameter and 15-25 cm in depth. Such holes may number about 100 to 120 in about 0.16 hectares.

The broodstock (25-30 cm long; 150-200 g each) are stocked in November/December in the holding channel at a stocking rate of 625 kg/ha. They are fed daily rice bran and trash fish in the ratio of 1 : 9 at the rate of 5-6 per cent of body weight.

Feeding is stopped a day or two before breeding, and water is pumped in to inundate the holes. Normally 25-35 cm of water is maintained above these holes. The broodfish are stimulated by the arrival of fresh water and move to the holes to spawn. Spawning takes place within one to two days.

Eggs hatch out within 18 to 20 hours. After hatching, the larvae remain huddled together in the holes for 9 to 10 days, after which they come out of the holes and start wandering. Therefore, they are caught using a hand scoop net while they are still in the holes. Then, the water level of the pond is lowered to bring the broodstock back into the holding channel. Feed is given for another 12 to 15 days before the operation starts again. This is done, on an average, about once a month. In October, the broodstock is sold out and a new batch is introduced into the holding tank.

Nursery ponds are 800-3,000 m^2 in size, with a 50 cm water depth. A channel 2 m wide and 50-60 cm deep is constructed in the middle of the pond to facilitate collection of fry during harvesting. Prior to stocking, chicken manure is applied to the pond at about 30 kg/100 m^2. Stocking rate is 250-350 fry/m^2. Fish are fed with ground trash fish and rice bran in the ratio of 9 : 1 at the rate of 1 kg of food/ 1,000 fry, twice a day. The fry reach 2-3 cm in 15 days. Survival rate is usually 30-50 per cent.

Grow-out ponds for the culture of *C. batrachus* are normally constructed in areas with heavy clay soils near a perennial source of water. The size of farms ranges from 0.1 to 1 hectare with pond size varying from 400 to 2,000 m^2. Each pond has its separate inlet and outlet channel. Preparation of ponds involves drying and occasional application of lime at 600-700 kg/ha. Two to three days before stocking, the ponds are filled with water to a depth of about 40-60 cm. Stocking rate is 100-300 fingerlings of 3-5 cm long/m^2. Ponds are stocked during February-April and harvested during May-August. For the first two weeks after stocking, the fish are fed with ground trash fish and a little rice bran. Subsequently a mix of trash fish, rice bran and boiled broken rice is made into a sticky paste. The proportion of rice bran in this feed mix is gradually increased as harvest time approaches, since rice bran helps in fattening the fish. Vitamin and mineral premix, and at times antibiotics, are sometimes added to this feed mix. Feeding rate varies from 120 to 400 $g/m^2/day$ during the culture period. Water is changed frequently. The fish are harvested after four to five months when they have grown to 100-350 g. On an average, the ponds yield 3-10 kg/m^2.

The technology for the culture of *C. batrachus* has also been developed in India. It is grown either on a monoculture basis or in combination with *Heteropneustes* and *Anabas*. As a substitute to common carp, the culture of *C. batrachus* with Indian and exotic carps in composite fish culture is also advocated.

Heteropneustes fossilis (Singhi)

Heteropneustes fossilis, commonly called stinging catfish is primarily a fish of ponds, ditches, swamps and marshes and tolerate slight brackish-water too. Apart from India, it is found in Pakistan, Bangladesh, Nepal, Myanmar (Burma), Sri Lanka, Indonesia, and Thailand. The maximum

Figure 13: *Heteropneustes fossilis* (Singhi)

recorded size of *H. fossilis* is 38 cm. Males are smaller than females. In one year, while females grow up to 12 cm, males grow to only 8 cm. As the name stinging catfish itself suggests, the fish is equipped with poisonous pectoral spines which can inflict severely painful wounds leading to local inflammation and even fever.

Sexual maturity in *H. fossilis* is attained at the end of the first year of its life. The spawning season extends form June to September. Fecundity varies form 3,000 to 45,000 according to the size of the fish. A female weighing 100 g, produces about 8,000 eggs on an average. Fertilized eggs are mildly adhesive in nature and are either green or brown in colour. The period of incubation varies from 18 to 24 hours depending on water temperature. Hatchlings measure 2.72 mm in length. Yolk gets completely absorbed by the end of the fourth day but the larvae commence feeding from the third day itself. Minute zooplanktonic organisms such as ciliates, rotifers, etc., serve as the choicest food of the larvae at this stage. Aerial respiration apparently commences from the eighth day of development. *H. fossilis* is an omnivorous fish and a bottom feeder. It has a great liking for gastropods.

H. fossilis is cultured both singly or in combination with *Clarias* and/or *Anabas*. Innovative devices are adopted to collect its young ones from nature to serve as stocking material for culture operations. Stocking material *of H. fossilis* can also be produced by breeding the fish through hypophysation and rearing the resultant spawn to stockable size.

In semi-intensive culture operations, the recommended rate of stocking for *H. fossilis* is 60,000 fingerlings/ha. When food is plentiful, the fish fattens well. Feeding with low-grade dried marine trash fish and rice bran has been reported to bring better growth. A mixture of oilcake, rice bran and biogas slurry in the ratio of 1 : 1 : 1 has been found to work as a good low-cost feed for *H. fossilis*. In monoculture, *H. fossilis* has yielded productions to the tune of 1,642 to 7,300 kg/ha in four to eleven months of culture period, the average being 4,510 kg/ha.

Ictalurus punctatus (American Catfish)

The most important aquaculture species of catfish in the U.S.A. is the channel catfish. It adapts readily to pond conditions as well as raceways, cages and tanks, accepts artificial feed, and tolerates crowded conditions associated with intensive culture. The fish is highly esteemed for its high quality meat, especially in the southern states of U.S.A., where the fish is considered as a traditional delicacy.

Other catfishes of cultural importance in the U.S.A. are the white catfish (*I. catus*) and the blue catfish (*I. furcatus*). Techniques for propagation have been developed so that large numbers of fingerlings can be produced for culture. Though brooders of channel catfish can be obtained form streams or lakes, farmers prefer cultured broodfish grown

Figure 14: *Ictalurus punctatus* (American Catfish)

on a diet rich in animal protein in a clean and healthy environment. Broodfish of 1 to 5 kg are used for breeding. Larger broodfish can also be used but the smaller ones are relatively easier to handle. Mature females ready to spawn have a well-rounded abdomen with the fullness extending past the pelvis to the genital orifice. The ovaries are palpable and soft and the genitals swollen and reddish. Males with a heavily muscled head wider than the body, dark pigmentation under the jaw and large protruding genital papilla, usually have well-developed testes.

Channel catfish fight during the spawning season. Their bites are often deep enough to damage skin. Fish sometimes die from severe bites. To avoid this, special care is taken while pairing the fish for breeding. Most culturists choose to use larger males. If the female is of a much larger in size, she usually attacks the male and does not mate with him.

Provision of a suitable nest is a major requirement for pond spawning. Shallow ponds of about 0.4 hectare usually serve well as spawning ponds but bigger ponds of up to 2 hectares are also sometimes used. Spawning ponds nests are made of nail kegs, wooden boxes, hollow logs, large milk cans, concrete tiles, metal drums, etc. The nests are usually placed in the marginal areas of the pond at depths varying from 15 to 150 cm, with the door of the nest facing the centre of the pond. The broodfish are introduced at the rate of 56 to 125 females per hectare.

Channel fish spawn in late spring and summer at a temperature range of 21 to 29°C. After introducing the broodfish in spawning ponds, the nests are checked regularly to see whether spawning has taken place. After spawning, the male guards the eggs during incubation and fans them with his fins to keep a current of well-oxygenated water flowing over them.

The period of incubation ranges form 10 days at 21ºC to five days at 29ºC. After the eggs hatch, the fry congregate at the bottom and remain there for about two days before coming to the surface. After hatching, the fry are collected (after driving away the male or removing it) and transferred to fry rearing trough or pond. Channel catfish is also bred in pens, aquaria and similar containers.

For fry rearing, trough method is preferred since it allows greater control and the fry are less exposed to predators. Rearing troughs are generally 2.5 to 3.0 m long, 30 cm deep and 20 to 50 cm wide. It is made of wood, metal, plastic or fiberglass. Each trough is supplied with running water. The fry from one or two spawns are put into each trough. The fry begin to feed shortly after yolk absorption and the fish begin to become greyish in colour. Commercial feeds containing 28 to 32 per cent protein and other nutrients are provided to the fry four to five times each day at the rate of 4 to 5 per cent body weight.

The size of the fry rearing pond is usually 0.4 hectare. Predatory insects are often a problem. They are removed by using non-residual insecticides two to three days before stocking. Fry are stocked at the rate of 50,000 to 250,000 per 0.4 hectare. They are fed daily along the marginal areas of the pond at the rate of 4-5 per cent body weight.

Rearing of fry to fingerling size is done in ponds or in trough. The ponds are usually 0.5 hectare in size. Stocked at the rate of 90,000 to 125,000/ha fingerlings of 15 to 20 cm are produced in about four months time. Starter feeds in meal or ground mash form may be changed to small pellets when the fish reach about 2.5 cm in length. The ration of feed is gradually increased from 0.5 kg per 0.8 hectare pond to about 4-5 per cent of the fry body weight.

Culture of channel catfish to market size takes a little less than two years after hatching, or one year from the fingerling stage. The usual market size is 500 g to 1.4 kg. For pond preparation, steps like eradication of weed by the application of rotenone, application of fertilizers etc. are taken. The stocking density depends upon quantity and quality of the water supply and the desired size of the market fish. In ponds with dependable water supply, a stocking density of 3,700 to 4,900/hectare is common. Use of commercially produced formulated feed is the most common practice. The most commonly used methods of feeding are hand feeding, self or demand feeding. Production in commercial pond farms normally averages around 1,500 kg/ha, although production of up to 3,000 kg/ha has also been reported.

Cage culture of channel catfish is generally carried out in open waters rather than in ponds. The usual rate of stocking is about 65 fingerling/m^3. Complete floating feeds are usually given at the rate of 3.5 per cent of body weight per day. This is gradually reduced to 2 per cent depending on the quantity consumed.

Chapter 14

Culture of Coldwater Fishes

Among coldwater fish, the trout's are esteemed owing to their sport value. They inhabit heavily in the rivers, stream, brooks, lakes and ponds of upland areas like Kashmir, Himachal Pradesh, Uttar Pradesh, Nilgiris, Kodai and hills.

Species Cultured

- ☆ Rainbow Trout: *Salmo gairdneri gairdneri*
- ☆ Brown Trout: *Salmo truta fario*
- ☆ Brook Trout: *Salvelinus fontinalis*
- ☆ Indian Trout: *Barilicus bola*
- ☆ Deep bodied mahseer: *Tor tor*
- ☆ Golden mahseer: *Tor putitora*
- ☆ Mosal mahseer: *Tor mosal*
- ☆ Atlantic salmon: *Salmo solar*
- ☆ Sockeye salmon: *Oncorhynchus nerka*
- ☆ Mirror carp: *Cyprium carpio var specularis*
- ☆ Leather carp: *Cyprium carpio var nudus*

Seed Resources

As coldwater fish prefer gravelly substratum to safe guard their eggs and the fertilized eggs stick to gravel and debris, it is easy to collect the eggs in large numbers from such a substratum.

Nursery Ponds

The nursery ponds may be concrete or stone walled with the size ranging from 2.5 × 1 × 0.75 m–9 × 1 × 0.75 m. These are provided with suitable inlets and outlets for free water flow and the water flow inside the nursery pond may be fixed at 100 lt./min. The stocking density should be 4-6 fry/m^2.

Rearing Pond and Raceways

Here the advanced fry are grown to fingerlings or adult. A series of rectangular raceways ranging 20-100 m^2 in size with a water depth of about 1.5 m may be constructed either as all concrete structure or concrete sides with earthen bottom. The bottom of every race way must be sloped so that it becomes shallower from inlet to outlet. They are connected with one another and are supplied by the water of streams and nearby rivers. Water flow should be 50 lt./sec.

Culture of Trouts

The trout species of greatest importance in aquaculture are the rainbow trout (*Salmo gairdneri*), the brown trout (*S. trutta*) and the brook trout (*Salvelinus fontinalis*).

Rainbow trout is native to North America. It has been introduced to waters of all continents except Antartica. Several strains of rainbow trout have been developed through mass selection and cross-breeding for improved cultural qualities.

Brown trout is indigenous to Central and Western Europe and was the first trout to be artificially propagated. It has also been introduced into several countries for developing sport fisheries. Because of its slower growth rate, poor utilization of artificial feeds and stringent water quality requirements, commercial cultivation of brown trout has not developed so much as rainbow trout.

Brook trout is native to northeastern North America. The species was introduced into Europe towards the end of the 18th century. But because of stringent water quality requirements and susceptibility to diseases, its culture has not been very common in spite of its fast growth rate and higher demand.

In Europe, the brooders of brown trout are caught from open waters when they start migrating upstream for spawning. Fast-growing strains with large eggs are now also available for culture. With the development of trout farming in sea water, interest in late maturing, fast-growing strains or sterile hybrids has increased. A recent development is the use of sex-reversed all female brood-stock for producing all-female progenies that grow faster.

Females are rarely used for breeding before they are three or four years old. Males of two to four years are considered ideal for breeding. The males and females are usually maintained in separate ponds. Maturity is examined at regular intervals so that as soon as fish is ripe it can be used for stripping. The methods of stripping, fertilization and egg incubation for different species of trout are basically the same.

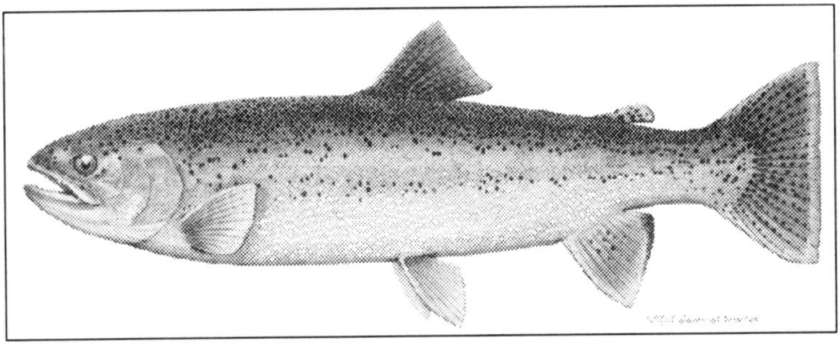

Figure 15: Rainbow Trout

Dry method of stripping is most commonly followed. The females are taken out and wiped gently with a towel, after which the eggs are stripped and collected in a dry bowl. Depending upon the quantity of eggs, two to four females are stripped at a time. Milt from males is stripped and poured on to the eggs evenly so as to cover as many eggs as possible. After proper mixing, the mixture of eggs and milt is left undisturbed for a minute or two following which water is poured in slowly to fill the bowl and it is left as such for 10 minutes. Now the eggs are given several washes in water to remove the surplus milt, and then they are transferred to bigger containers.

For the incubation of eggs the system of troughs and baskets are most common. The troughs are usually 40 to 50 cm wide and 20 cm deep with the length varying to about 3 to 4 m. Rectangular baskets are placed in these troughs, with their top edges sitting on the sides of the trough. The perforated bottom of the basket rests about 3 cm above the bottom of the trough. The perforations are such that they retain the spherical eggs but allow the hatchings to fall through to the water below. The baskets are arranged in such a way that the water keeps passing through the mesh to aerate the eggs. The eggs are spread one or two layers thick. The water flows from the top end of the trough and the eggs flows on to the next basket, until it reaches the end of the trough. An average-sized trough needs a continuous water flow of 5,000 l/day for every 10,000 eggs. When a large number of eggs have to be incubated and where there are space limitations in the hatchery, it may be necessary to use battery incubators.

The period of incubation varies form 21 days at 14.4°C to 100 days at 3.9°C. From the "eyed" stage (when the eyes can be seen through the egg shell) until hatching, the eggs are quite tough and withstand handling and transport rather easily. The hatchlings remain in the hatchery baskets until they reach the swim-up stage and all the yolk has been absorbed.

Fry are reared in hatchery trays itself for some days, after removal of the egg baskets. However, when battery incubators are used, it is essential to transfer them to a rearing tank. Fry rearing is carried out in indoor concrete or fiberglass tanks. It is easier to maintain a regular flow of water in such tanks to facilitate uniform distribution of fry inside the tank.

In modern trout hatcheries, specially prepared starter feeds are given to the fry, using automatic feed dispensers. In the beginning, the fry are fed five to six times a day and as they grow older, the frequency of feeding is reduced to twice a day. When commercial feed is not available, conventional feeds consisting of meat, liver and fish, finely ground to a size that can be swallowed by the fry is provided. Fry rearing, normally lasts for 10 to 12 weeks.

Earthen ponds excavated in ground level, with water from a nearby river or other suitable water source, supplied by gravity, are commonly used for farming trouts. Farm ponds are rectangular in shape, usually about 30 × 10 m with the bottom sloping towards the outlet. The depth at the upper end is 1 m and 1.7 m at the lower end. The capacity of the ponds is about 1,500 kg of fish.

Trouts are farmed in many parts of the U.S.A. mainly for recreational fishing and to provide an additional crop of highly relished food to the farmers. Trout culture in raceways is the most widely used system. This is done mainly with the view of rearing trout for restocking sport waters. Raceways consist of long continuous channel or a series of channels divided by cross-walls. Plentiful supply of clean water, flowing through the channels is the most important requirement of raceway culture. A stocking density of 4-5 kg/m^3 is used when there is water exchange of 2.5 l/minute/m^3.

In Europe, trouts are cultured in tanks of 4-10 m diameter with a depth of about 1.6 m. Tanks are generally sunk in the ground, leaving a portion of about 30 cm above the ground. The drain pipes are laid in trenches. Cage farming of trout is also rapidly expanding. Monoculture is the most common practice in trout culture. In some areas, double cropping systems have also been introduced. The double cropping of trout and channel catfish is a common practice in southern U.S.A.

Culture of Salmons

The most important species of salmon are Pacific or King salmon (*Oncorhynchus tshawytscha*), Coho salmon (*O. teisuthc*), Sockeye salmon (*O. nerka*), Chum or Dog salmon

(*O. keta*), Pink or Humpback salmon (*O. gorbuscha*), and Atlantic salmon (*Salmo salar*). They are mostly utilized for ranching programmes.

King salmon is the largest species growing up to 45 kg in weight. It is widely reared in fresh water for enhancement of depleted stocks in North America. Maturity and spawning of the fish usually occur between two and seven years of age. Coho salmon is a hardy species and has been widely transplanted. It attains a weight of 15 kg, maturing at an age of two to four years and migrating upstream in late fall or winter. It is regarded as a preferred species of Pacific salmon for farming in Japan, the U.S.A. and Canada.

The native of sockeye salmon is east and west Pacific. The greatest numbers are found from Bristol Bay to the Columbia river on the east, Kamchatka on the west. Though it grows up to 100 cm in length, the average size of sockeye salmon is much lower. Its weight at maturity varies from 1 to 5 kg. It is disease prone in captivity. Chum or dog salmon is native to east and west Pacific. It enjoys the widest distribution among all the Pacific salmon species. It grows to 1 m in length and 20 kg in weight. Humpback salmon is native to east and west Pacific. It grows to 4 to 5 kg. The size attained by the Atlantic salmon in nature varies greatly, depending upon the length of its ocean residence but weights up to 36 kg have been recorded. The usual weight of Atlantic salmon is 2 to 10 kg averaging about 4.5 kg.

For artificial propagation, broodfish are held in sea or fresh water for stripping but sea-grown fish are slowly acclimatized if they have to be transferred to fresh water. Mature male salmons are recognized by their hooked lower jaw. Dry method of stripping is used for fertilization. The

Figure 16: Coho Salmon

stripped eggs and milt of salmons are sensitive to light so they are kept away from direct sunlight. Since Pacific salmons die after spawning, the eggs are taken out by cutting the fish. The milt is generally stripped from the males, so they are not killed. The trough and battery type incubators are commonly used for hatching.

Salmon fry can be reared in high densities of 10,000 fry/m^2 tank surface. Subsequently, however, the density is reduced, the water flow in the tank is maintained at the rate of 1 litre/minute per kg fish. The smoltification of salmon takes place in spring when it undergoes physiological pre-adaptation for life in sea water. The minimum size at smolting is about 15 g, but the desired size at the end of the first year is about 30 to 50 g. Larger size smolts are preferred for grow-out.

The most common method of culturing salmon to market size is floating sea cages or pens. Besides adequate feeding and frequent monitoring of the oxygen concentration of the water, the health of the fish is required to be regularly checked.

Chapter 15

Culture of Common Carp

Cyprinus carpio, popularly called common carp, European carp or German carp is a native to China and Russia from where it has been transplanted to Europe, Asia and other parts of the world. At present, common carp enjoys global distribution occurring in the tropical as well as temperate regions where it has acclimatized itself to a variety of habitats and extremes of environment.

Three phenotypes of common carp have been recognized based on the pattern of their scaling:

- ☆ The scale carp, *Cyprinus carpio var. communis* with its body fully covered by regularly arranged rows of scales. This is the original form of common carp with normal colouration.
- ☆ The mirror carp, *Cyprinus carpio var. specularis* with its body covered, unevenly with a few large and bright scales. A large area of the body, however, remains naked. The species is extremely variable and generally look orangish in colour.
- ☆ The leather carp, *Cyprinus carpio var. nudus* with its body almost completely devoid of scales except

for a single row of somewhat degenerated scales along the base of the dorsal fin and sometimes extending as far as the tail or even from head to tail. Sometimes, small scales are also present at the bases of other fins.

Common carp is an omnivorous fish. It feeds on a wide variety of plant and animal matters. While young fish of less than 10 cm length feeds on zooplanktonic organisms, the bigger fishes feed on benthic organisms such as insect larvae, worms and molluscs, together with heavy intake of vegetable matter and epiphytic organisms. Its habit of sucking food matters in the mud on pond bottom and margins makes the water muddy and weakens the base of pond embankments.

Common carp spawns generally in confined waters of ponds. The spawning season for common carp in temperate climates is in the spring, when water temperature rises above 18°C. However, by manipulating environmental conditions and the use of selected races or strains, it is possible to extend its spawning season to suit the requirement of culture.

Generally, two-year-old fish weighing 2 to 3 kg are selected for breeding. Larger fishes pose problems in handling. About 100,000 to 150,000 eggs are produced per kg body weight of the fish. Mature females have a plump body with swollen abdomen while males look slender and ooze milt freely when gently pressed at the abdomen. It is desirable to segregate the males and females and maintain them in separate ponds to avoid unwanted spawning. Brooders are maintained on a protein rich diet to promote development of gonads. High carbohydrate feeds are avoided to prevent accumulation of fat in the body of the fish.

Common carp is bred in two ways: one, to breed naturally under the influence of simulated natural spawning habitat without giving hormone injection, and two, by resorting to hypophysation.

For the inducement of natural spawning, mature spawners are selected and released in prepared ponds or in cloth "hapa" fixed in a pond. As common carp lays sticky eggs, they are required to be taken care of by providing proper substrate for the eggs to adhere upon. Different types of substrates are used. In Indonesia, spawning mats made of fibres of "indjuk" (*Arenga* sp.) called "kakabans" are used. They are placed on long bamboo poles held in place between two pairs of shorter poles driven into the bottom. Israeli farmers use branches of pine, casuarinas and cypress as egg collectors. In India, aquatic weeds like *Hydrilla* is most commonly used as egg collector inside cloth hapas. Often, spawning substrate made of banana leaves, synthetic fibre or water hyacinth is used.

Ripe females (2 to 5 kg weight) are used at the rate of 10 nos/ha and the ratio of females to males is about 2 : 3. Fishes generally spawn in the early morning hours after their release in spawning ponds/tanks/hapa. Released eggs can be clearly seen sticking to the spawning mat/weeds. The period of incubation is 40 hours at 28 to 31°C.

The technique of hypophysation is generally adopted when conditions are not favourable for successful natural spawning in ponds. For hypophysation, selected brooders are used directly. Under normal conditions, a good female produces about 150,000 eggs/kg. The eggs measure 0.9 go

1.6 mm in diameter. Fish pituitary gland extract is prepared following the usually recommended procedure. The dosage and intervals of injections, however, vary from place to place. The injection is given intramuscularly between the base of the dorsal fin and the lateral line. The injected males and females are kept in separate tanks.

At first, the female is stripped by gently squeezing the abdomen towards the tail. The eggs are collected in a plastic tray, bowl, basin or trough. A good female may yield ripe eggs to the extent of about 20 per cent of its own body weight. A 5 kg female for example produces about 1 kg eggs, containing 700,000 to 1,000,000 number of eggs. Following the stripping of females, males are stripped in a similar manner and the oozing milt is poured on to the eggs. A 4-5 kg male may provide an average of 20 ml of milt.

Eggs and milt are mixed immediately using a feather or a plastic spoon. The adhesive nature of eggs makes them clump together and hamper proper fertilization. The stickiness can be removed by treating the eggs with a solution of sodium chloride and carbamide (40 g sodium chloride and 30 g carbamide dissolved in 10 litres of water). The solution is first poured over the mixture of eggs and milt and stirred with a plastic spoon or feather for about 5 to 10 minutes. As the eggs begin to swell, small quantities of the solution are added at intervals, as per need. The eggs are then washed in tannic acid solution (0.05-0.07 per cent) for about 20 seconds. The washing is repeated up to five times, the solution being diluted each time by about 0.01 per cent by the addition of water to the stock solution. Finally, the eggs are washed in fresh water for about five minutes. The water-hardened eggs, which measure 2 to 2.1 mm in diameter, are ready for incubation in hatchery jars or other suitable containers.

Different types of incubators are used for hatching common carp eggs, ranging from double-walled cloth hapa to zong jars with temperature-controlled water supply. Funnel incubators are also used widely. The eggs hatch out in two to three days, depending upon the temperature. The hatchlings are removed to indoor nursery tanks. It is desirable to rear the hatchlings under controlled conditions up to the fry stage. About a million fry can be reared in a $20m^2$ tank with a water exchange rate of one litre per minute per square metre. Best results are obtained when larvae are fed on natural food collected from ponds or other water bodies or cultured in farm.

For the nursery treatment of the fry produced in a hatchery, specially prepared nursery ponds of about $2{,}000m^2$ areas are most commonly used. In such ponds, the fry is raised to fingerlings size within a period of about one month. The management practices of nursery ponds are the same as for the Chinese and Indian major carps.

In India, common carp is usually cultured in combination with Indian carps. The growth rate differs greatly in accordance with the environmental factors, such as, temperature, availability of food, stocking density, supply of artificial feed etc. Polyculture of common carp together with tilapia and other species like tawes (*Puntius javanicus*), kissing gouramy (*Helostoma temmincki*) and giant gouramy (*Osphornemus goramy*) is practiced in southeast Asian countries, particularly Indonesia. In Poland and Czeckoslovakia, common carp is grown in combination with rainbow trout (*Salmo gairdneri*).

In majority of the countries where the culture of common carp is practiced, the market demand is for a fish of above 500 g, preferably 1 to 1.5 kg except in some Southeast Asian countries where smaller-sized fish are acceptable. The required marketable size in many countries, however, it usually takes two to three years to grow common carp to the preferred weight of 1 to 1.5 kg.

Common carp has been one of the most versatile fish cultured globally and its seed production and culture methods are easier than those of either Indian or Chinese carps. It is most promising for developing countries which do not have much of a fish-culture history or background, to initiate the culture of common carp.

Chapter 16

Culture of Gouramis and Climbing Perch

Gouramis and climbing perch are different from other fishes because of a unique accessory air-breathing organ, the labyrinth, located on either side of the gill chamber. The organ enables these fishes to survive in waters with very low dissolved oxygen.

Gouramis

Three species of gouramis are important for culture. They are the giant gourami (*Osphronemus goramy*), the Siamese gourami or sepat siam (*Trichogaster pectoralis*) and the kissing gourami (*Helostoma temmencki*).

The giant gourami (*O. goramy*) is the most highly priced species and has been introduced into a number of countries in the Indo-Pacific region. It grows up to 61 cm in length. The young of the species feed on planktonic organism but adults show preference for aquatic vegetation. The optimum temperature requirement of giant gourami is 24 to 28°C.

In nature, the giant gourami spawns during the dry season but in ponds, throughout the year. It breeds by

Figure 17: Giant Gourami

making submerged nests made of plant material. Spawners are generally two to three-year-old and both female and male guard the progeny. A five-year-old female produces 3,000 to 5,000 fry per year in two to three spawning at bimonthly intervals.

The simplest system of breeding adopted is the provision of suitable water plants or other material for nest making. Sufficient number of males and females in the sex ratio of 1: 2 or 2: 3 are kept in the pond. It usually takes about 10 days to make the nests. Spawning takes place two or three days after the nests are ready. A characteristic fishy smell and an oily substance that emanates from the nest indicate that spawning has taken place. The eggs hatch out and hatchlings become clearly visible in two to three days. While ants are considered to be the best food for larvae, peanut waste is sometimes used as a substitute. For further rearing, fry are usually fed on *Azolla* or minced leaves of other plants.

Giant gourami can be grown in fresh and slightly brackish water. It is a slow-growing fish. The fish usually takes two to three years to reach marketable size. The feeds used in grow-out ponds are leaves and tender parts of *Ipomea*, land grass, yam, tapioca, kitchen waste or a mixture of cooked rice, rice bran and trash fish. The rate of growth in the first year is 15 cm, 25 cm in the second year and 30 cm in the third year. The production usually averages to 200 kg/ha.

The Siamese gourami, also called snake-skin gourami (*T. pectoralis*), grows up to 25 cm in length and 100 g in weight. Its cage culture is very popular in Vietnam, Thailand, Malaysia and Kampuchea. It spawns in floating bubble nests which are made by the male in 70 to 100 cm deep spawning ponds. Spawners are usually about seven months old and about 100 g in weight. Fecundity varies from 20,000 to 40,000 in the fishes of 90 to 120 g weight range. The spawning nest which consists of a mass of bubbles, takes about two days to make and when ready, the female lays the eggs under it which are immediately fertilized by the male and tucked into the nest. Hatching takes place in one to three days. About 4,000 fry are obtained from each nest. The species also breeds easily in well-oxygenated shallow ponds containing emergent weeds to protect the nests.

T. pectoralis can tolerate low salinities and is raised in abandoned paddy fields inundated by sea water. In the traditional method of culture, the paddy fields are modified as per the cultural requirements of fish and used as culture ponds. The bunds are widened and peripheral trench 3 m wide and 0.75 m deep is made. A central platform is maintained to provide space for growing grasses and weeds which are periodically cut to fertilize the ponds.

For breeding, breeders of 10-12 cm length are released at the rate of 200 pairs/ha. A female normally produces as many as 7,000 to 10,000 eggs of which 300 to 500 survive to fingerling size. Pond management comprises steps like weekly manuring with grasses cut from the peripheral dikes and central platform, periodical addition of water to compensate for evaporation and seepage losses. The fish reaches a marketable size of about 100 g after eight months. The average yield of 1,000 kg/ha/yr is obtained.

A notable innovation in the improved culture technique of *T. pectoralis* is the construction of a breeding pond which is distinct and separate form a grow-out pond. By this innovation, the mortality rate of fry and fingerlings is considerably reduced. The breeding pond is usually a small enclosure fenced with nylon netting to prevent entry of predators. Besides, improved method of culture, fertilization of the pond with organic and inorganic fertilizers is also done which enhances the growth of natural food. The production under improved method of culture ranges form 1,000 to 2,000 kg/ha/yr.

The kissing gourami (*H. temmencki*) derives its name form its habit of characteristic labial contacts with other individuals of its species, suggesting as if they are kissing each other. It is a native to Thailand, Vietnam, the Malay Peninsula and the Western islands of the East Indies. It reaches a length of about 30 cm and matures at 20 cm or at 12 to 18 months of age. It is a fresh water fish of predominantly plankton feeding habit but it accepts almost all sorts of artificial feed.

The kissing gourami is bred either in separate spawning ponds or harvested in paddy fields. In Indonesia, the spawning ponds vary in size form 30 to 1000 m in area with facilities for clear water supply. Rice straw or banana leaves are spread on the pond to provide shade for the eggs and to protect the larvae. Spawners are held in a segregation pond for a period of about one month after which they are released into spawning ponds. Spawning takes place in about 18 hours and the larvae hatch out in about two days. The ponds are manured with decaying plant material about seven to 10 days after the larvae have hatched out in order to develop enough of planktonic food for the larvae.

Climbing Perch

The climbing perch *Anabas testudineus* is a fish of tropical fresh waters of Asia and the Far East. The fish derives its name from the widely held belief that it can climb on trees, which is actually a result of faulty observation. It is true that sometimes *Anabas* is found lying in the forked branches of trees or inside toddy containers hung on palm trees but it does not reach there by itself. It is placed there by kites and crows that pick it up from drying pools or during its over-land travels of migrating from one water to another.

A. testudineus is a very hardy fish. It is mostly caught from swampy and marshy tracts as well as derelict pits, pools and puddles which remain in the process of drying up during summer months. The fish in its natural habitat can virtually "walk" over day land for great distances at a stretch using its operculum and pectoral fins.

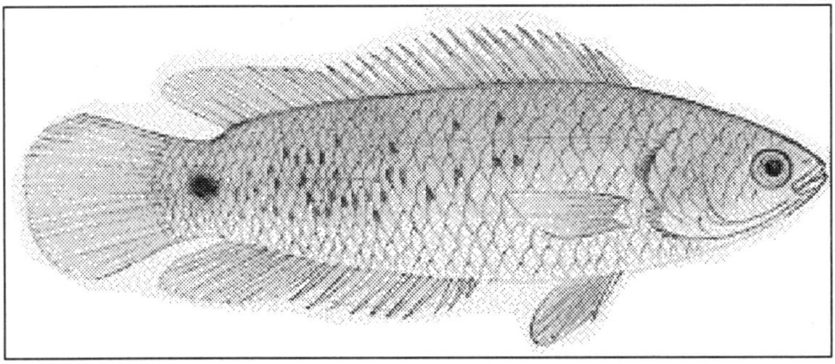

Figure 18: Climbing Perch

Despite its moderate size, *A. testudineus* is a highly esteemed fish for its fine flavour. The maximum recorded size of the fish is 26 cm. Gonadal maturity in *A. testudineus* manifests with early rains and the retrogression is noticed abruptly form September. Breeding takes place only after the rain fully sets in and fills the ditches around. The fecundity of the fish ranges form 5,000 to 35,000 depending upon its size.

Seed of *A. testudineus* can be collected from nature and also obtained under controlled conditions. *A. testudineus* is notorious for migrating and breeding into prepared carp nurseries. This habit of *A. testudineus* is advantageously made use of for its seed production. A female weighing about 50 g produces about 20,000 eggs. Eggs float freely and look yellowish in colour. The period of incubation varies form 18 to 24 hours at 27 to 30ºC. Hatchlings, measuring 1.9 mm in length, float in an upside-down position at the surface. Feeding commences from the second day itself. Ciliates, flagellates and other such minute zooplanktonic organisms form the chief food of larvae at this stage.

A. testudineus is an omnivorous fish but shows definite preference for insects. The scope of its natural food, however, is very wide. It can vary from a diet of filamentous algae to pure carnivores.

A. testudineus can be cultured singly or in combination with *C. batrachus* and/or *H. fossilis*. It can also be grown in composite fish culture ponds together with carps. *A. testudineus* has a tendency to "walk out" from the ponds during rains. It is, therefore, necessary to select only such ponds for its culture in which the dykes are sufficiently steep (75° or more) so as to prevent the fish from escaping out. Depending on the management measures, the rate of stocking for monoculture operations may vary from 25,000 to 50,000 ones/ha. The measures of management are basically the same as in *C. batrachus* and/or *H. fossilis*. In monoculture, average production to the tune of 876 kg/ha in five to 11 months of culture period has been obtained in India.

Chapter 17

Culture of Giant Freshwater Prawn

Giant freshwater prawn, *Macrobrachium rosenbergii* (de Man) is fast becoming one of the most important cultured species in the inland aquaculture system in the country due to its fast growth rate, high market demand, attractive price and its compatibility with carps. This species can also be cultured in low saline brackish water areas (salinity<7 ppt) and can be cultured either alone (mono culture) or in combination with carps, tilapia and *Chanos* (poly-culture). It is also a suitable species for incorporation in the rice fish system of culture. Culture of *M. rosenbergii* can be carried out in earthen ponds, cement cisterns, in pens or in cages. However, most of the operations are being carried out in earthen ponds.

Site Selection

In case of farmers who already own a pond, the culture can be carried out in the existing ponds. For those who wish to start it fresh then proper site selection is essential to obtain best results. The selected site should have the following major qualities:

Figure 19: Freshwater Prawn

- ☆ Supply of good quality, pollution free freshwater or brackish water (<7 ppt) for at least 8 months
- ☆ Soil having a pH of more than 6.0 and good water retention capacity.
- ☆ Warm climate for nearly 8 months (Temperature - > 24°C).

1. Pond Construction

Regarding construction, the ponds should be preferably embankment ponds that can be fully drained by gravity. Important points to be remembered while construction of ponds are provided below:

- ☆ Ponds should have an inlet and an outlet.
- ☆ Pond bottom should have a gradient slope towards the outlet (1 : 200).
- ☆ Pond bunds should have a suitable slope (1 : 2, 1 : 3).

- ☆ A water control structure should be installed (wooden/concrete) to aid water exchange.
- ☆ Pond size—0.2 -<2 ha (0.2-0.5 ha).
- ☆ Shape—Rectangular with their long axis oriented in the direction of prevailing wind is most suitable.
- ☆ Soil—Clay loam, sandy loam.
- ☆ Depth—2 m.

2. Water Quality

Water used for culture should be free from toxic chemicals and pollutants. The optimum ranges of few most important water quality parameters for freshwater prawn culture are given below:

- ☆ Salinity: Freshwater/low saline (<7 ppt)
- ☆ Temperature: 28–31°C
- ☆ pH: 7.0–8.5
- ☆ Total hardness: 40–100 mg/l
- ☆ Dissolved oxygen: > 4ppm
- ☆ Calcium: 50–100 mg/l

Pond Management

The important management practices needed for pond preparation, feeding, water quality measurement, prawn sampling and finally harvesting.

1. Pond Preparation

The steps involved in pond preparation includes, eradication of predators and competitors, application of lime and application of fertilizers (organic and inorganic).

(*a*) Eradication of Predators and Competitors

In a newly constructed pond here may be nothing to eradicate. But in old ponds all the unwanted species, including aquatic vegetation should be removed. Eradication of predators and competitors can best carried out by drying the pond and exposing the bottom to sunlight until cracks develop. Those ponds that cannot be drained should be applied with poison to kill all predatory fishes and other animals. Poisons of plant origin such as mahua oil cake, tea seed cake or derris root powder are preferred as they also will act as fertilizer after a period of time. Bleaching powder and urea @ 300 and 100 kg/ha respectively can also be applied to kill all predatory species in the pond. For this, first urea and after six hours bleaching powder has to be applied. Stocking with prawn seed can be done two weeks after eradication of pests and predators.

(*b*) Application of Lime

Lime is applied @ 200-1000 kg/ha depending on the soil pH. Lime helps to correct the pH; increases the buffering capacity of the water and disinfects the pond bottom. Lime is also a source of calcium, which is very important for exoskeleton formation of prawns.

(*c*) Application of Fertilizers (Organic and Inorganic)

Fertilizers both organic and inorganic can be used. Manure helps in the development of phytoplankton cover, which in turn prevents the development of benthic algae and rooted vegetation in the ponds. Cow dung @ 1000-2000 kg/ha or poultry manure @ 500 kg/ha and super phosphate @ 100 kg/ha may be applied to initiate a plankton bloom. After the initial fertilization water can be filled up to the desired level (4-5 feet).

2. Provision of Hide Out

Prawns exhibit territorialistic behaviour and are pugnacious in nature. They also need shelter during moulting to avoid predation by other prawns. Hence it is desirable to provide hideouts and more surfaces for clinging. Cut branches of trees, nylon screen, earthen pipes etc. can be used as hideouts.

3. Stocking the Pond

After preparing the ponds and laying the hide outs the ponds can be stocked with post larvae or juveniles as the case may be. Prior to stocking the pond water quality should be tested and necessary correction should be made. For stocking, seed can be collected from natural collections or from hatchery. Natural seed may contain unwanted species and its availability is restricted.

On arrival to the farm site care should be taken to acclimatize the post larvae to the temperature of the pond by floating the transport bags in the ponds for 15 min. After opening the bag some pond water should be allowed into the container and slowly the post larvae should be released in to the pond. Early morning and late evening are considered ideal period for stocking the seed. The density of stocking depends on the market size desired and intensity of management. For semi-intensive culture a stocking rate of 5-8/m^2 is desirable. In poly-culture stocking density of prawn is reduced to 50 per cent and compatible carp species such catla, rohu, silver carp and grass carp are also stocked @ 3000/ha.

4. Food and Feeding

Freshwater prawns are omnivorous, and should be fed with both vegetable and animal matter. Successful commercial farming of freshwater prawns must involve supplementary feeding. Pellet diets are preferred to maintain the water quality in the ponds. Experiments have shown that these prawns require 25-30 per cent protein and 8-10 per cent fat in their diet. Feed should be spread evenly along the peripheral area of pond. Use of check trays placed at different locations in the pond will be helpful to check the consumption rate and for managing the feeding rate. Since the prawns are more active during night, feeding should be done during late evening and early morning. Daily observation regarding the consumption of feed should be the main criteria for deciding the quantum of feeding. The feeding rate of prawns is also related to water temperature. At low water temperature (<24°C) the feeding rate drops to less than half and at less than 20°C they almost stop feeding. Monthly sampling should be done using cast net to know the average body weight of prawn to adjust the feed quantity.

Table 12: Weight Dependent Feeding Rates

Prawn Weight (g)	*Percentage of Body Weight Fed Daily*
<1	>25
1-2	20
2-5	15
5-10	10
10-15	8
15-20	7
20-25	6
25-30	5
>30	2

5. Water Quality Management

Daily monitoring of critical water quality parameters such as dissolved oxygen, pH, and temperature is essential to prevent any loss of stock due to poor water quality. Dissolved oxygen should be monitored during early morning. On cloudy days and rainy days depletion of oxygen may occur during daytime also. Phytoplankton bloom and decaying waste material are the main reasons for dissolved oxygen depletion usually seen in prawn ponds. When the oxygen level in pond water is critically low then prawns come to the surface along the periphery of the pond. Immediate remedial actions such as water exchange (with good quality water) or operation of pond aerators should be taken to avoid mortality of stock.

Visibility and colour of the pond water gives a visual assessment of the condition of the pond ecosystem. In unproductive ponds the visibility can be up to the bottom. In highly blooming and or turbid ponds the visibility will be up to a few centimeters (<10 cm). Visibility should be maintained in the range of 30-40 cm to avoid water quality deterioration

6. Stock Monitoring and Estimation

Growth of the animal is assessed by regular monthly sampling with cast nets or small mesh seine nets. The growth rate and survival of a population of prawn depends on many factors including density, predation, feed and temperature. Since these factors are highly site and operator specific it is very difficult to predict the growth rate.

7. Disease Control

At present diseases have always been found to be linked to poor rearing conditions (over feeding, water shortage, silting). Preventive measures are given below:

- ✩ Maintaining a good rearing practice, avoid high density stocking and over feeding;
- ✩ Drying out the ponds between production cycles so that the beds can be reoxidized;
- ✩ Water exchange (renewing 30-50 per cent of the pond volume in 24 h) that rinses the pond and induces moulting.

Protozoan parasites, bacteria and fungi cause diseases. Loss of appendages, brown or black colouration of the exoskeleton, etc. can be seen in diseases affected prawns. They may not accept feed. If disease symptoms are noted water should be replaced. Water quality should be tested to determine the DO; pH and ammonia levels and necessary corrections should be made.

8. Harvesting

Because of the heterogenous growth pattern of *Macrobrachium rosenbergii*, individual weight is highly variable for prawns of the same age. Periodic harvesting is always preferred. After four months, bigger size prawns (>30g) can be removed by using a seine net of suitable mesh size. Selective harvesting should continue once every 3-4 weeks for another 3-4 months and finally the pond may be harvested by complete draining.

9. Survival, Growth and Yield

An average survival of 60 per cent can be expected from a properly prepared pond, stocked with quality seed, fed with quality feed and monitored regularly to maintain optimum water quality. The average expected body weight of prawns after 6-8 months of culture is 40g. Yield may range from 1-1.5 tonnes/ha/6-8 months.

Chapter 18

Freshwater Pearl Culture

Before 4000 BC pearls from the mid-east were considered to be the symbol of mystery, purity and perfection. It is the Chinese who introduced pearl culture in 1200 AD in Taihu, a sweet water lake in Central China. The Chinese are world leaders in culture of freshwater pearls. The first marketable crop was produced in 1970. There are more than a thousand freshwater pearl farms in China.

Freshwater pearl of superior quality were first produced in Japan, especially from its freshwater lake Biwa. In recent years, lake Keshumiga, Ura the largest lake in Japan is the most important source for freshwater cultured pearls. The technology is however firming its foot in Australia, America and different Asian and European countries like Indonesia, French Polynesia, Taiwan, Cooks Island, Malaysia, Philippines, Vietnam, Bangladesh in recent years. In India freshwater pearl culture is not even a decade old. It is estimated that pearl traders in India import around US$0.2 billion of pearls every year from Japan and China to meet growing domestic demand. India imports pearl and re-exports them after processing in the country.

What is Pearl?

- ☆ The pearl is a biological gem.
- ☆ Pearls are among the oldest and the most universal gems, being admired and adored as symbols of beauty, wealth and love the world over.
- ☆ In nature, a pearl is borne of a "Love-hate" relationship between the shellfish and foreign body that invades its shell.
- ☆ If the intruder cannot be ejected out, the shellfish "lovingly" encase it with a precious nacreous substance and make a pearl out of it.
- ☆ Unlike the precious metals that needed to be facetted and polished, pearls naturally arrived in a beautiful form that people could use.
- ☆ To meet the rising demand for pearls in the modern world, entrepreneurs and researchers resorted to pearl farming.
- ☆ However, the "farming" itself needed better yields and superior quality, and hence the need for innovative technology arose.

Structure of Pearl:

- ☆ Structure of pearl is similar to that of the inner nacreous layer of the shell.
- ☆ Chemical it is calcium carbonate, but the layer orientation and composition of aragonite crystal layers in nacreous pearls give the gem quality.
- ☆ There are five characteristics to consider in determining a pearls value, they are

- o Texture
- o Shape
- o Colour
- o Size and
- o Luster

Types of Pearls

There are essentially three types of pearls:

1. Natural
2. Cultured and
3. Imitation

- ☆ A cultured pearl undergoes the same process as that of natural pearl, the only difference being that the irritant which is a surgically implanted round shell bead (nucleus) or piece of shell called mother of pearl.
- ☆ For imitation pearls, a glass bead is dipped into a solution made from fish scales for acquiring luster.
- ☆ In this method, however, the coating formed may eventually grow fainter.

Composition of Pearls

- ☆ A pearl is a biological gem having colour and cool luster among precious gems.
- ☆ The composition of pearl is about 47 per cent aragonite crystals of calcium carbonate, 50 per cent organic matrix–a scleroprotein called conchoien and 3 per cent water.
- ☆ It has 4.0 hardness of Moh's scale with a specific gravity of 2.7.

- ☆ In a way, a pearl is an extension of the inner strong layer called nacre of pearl oyster in the marine and certain mussel species of in freshwater environment.

The freshwater mussels used for pearl production in India are:

1. *Lamellidens marginalis*
2. *Lamellidens corrianus*
3. *Perreyua corrugata*

- ☆ Among the three species, *L.marginalis* crosses all others in qualitative importance for being used for the production of pearls.
- ☆ This species is available in almost all types of inland freshwaters.

Systematic Position of Freshwater Mussel, *L. marginalis*

Phylum	:	Mollusca
Class	:	Bivalvia
Order	:	Unionida
Family	:	Unionidae
Sub-family	:	Unioninae
Genus	:	Lamellidens
Species	:	Marginalis

The Identifying Features of Freshwater Mussel (*L. marginalis*)

- ☆ The shell of this species is belong to ovate, thin and smooth.

Figure 20: Freshwater Mussel

- ☆ The colour of the periostracum is blackish brown, shining with a light brown border along the ventral margin.
- ☆ Posterior end of the shell is broad, roundly angular with the wings grown narrow.
- ☆ Dorsal margin is slightly curved, anterior side short and narrow, the ventral margin contracted in the middle.
- ☆ Hinge has two cardinal teeth, the lower one more developed.

Habit and Habitat

- ☆ Freshwater mussels are commonly found in the freshwater lakes, rivers and ponds in India, Srilanka, Bangladesh and Burma and in South Asia.
- ☆ They usually burrow in the bottom mud and perform locomotion with the help of their large ventral foot.

Food and Feeding Habit

- ☆ *L. marginalis* is planktophagus (Patil 1979).
- ☆ Their principle natural food consists of diatoms, spores of fungi, microscopic algae and other minute organisms.
- ☆ The food particles are filtered from the water current that passes through inhalent siphon and get expelled through exhalent siphon.

Mussel Culture

The culture technique consists of:

1. Selection of rearing pond
2. Pond preparation
3. Liberation of mussel
4. Periodical manuring
5. Living and
6. Monitoring of water quality

Pearl Culture

· For pearl culture, healthy mussels are collected by hand picking from the bottom of natural water bodies like lakes, rivers, ponds etc., during February–April. These collected mussels are transported to the culture site with the temporary maintenance of 20-30°C uniformly covered with moist aquatic weeds to prevent dehydration.

- ☆ Acclimatization is done in collection tanks.
- ☆ There are three basic management steps involved:
    1. Pre-operative
    2. Operative and
    3. Post-operative management

Pre-operative Management

- ☆ This is the most vital step and any error in the selection of right type of mussels will lead to abortive exercise. The criteria diagnosed for selection of mussels are:
- ☆ Mussels are selected a day before implantation. These mussels are sorted into two groups, mussels to be operated are termed as "recipient mussels" and those to be sacrificed for graft preparation are termed as "Donor mussels".
  - (*a*) *Donor*: Healthy injury-free mussels of the age group 2-4 years and of length varying from 9-16 cm.
  - (*b*) *Recipient*: Healthy, injury and disease free mussels of the age group of 2-5 years and of length varying from 9-12 cm.
- ☆ The healthy mussels must have to satisfy the following pre-requisite:
  - (*a*) It should have fully developed gills and strong foot.
  - (*b*) The pallium should be free of injury on visual moderation.
  - (*c*) Strong abductor muscle to resist any attempt to force open the shell values and
  - (*d*) When subjected to any shock by applying external stimuli, the shell values shall close instantly, rejecting a jet of water.

Operative Management

- ☆ Operative management consist of implanting foreign particles preferably cut pieces of mantle epithelium and conch shell beads the ideal season for implantation is considered to be February-April.

- ☆ Ideal temperature range is 20-30°C.
- ☆ The materials required for implantation are:
  - (*a*) The graft
  - (*b*) The nuclear bead

(*a*) The Graft

The graft tissue is nothing but the mantle epithelium cut into small pieces of 2-3 cm^2 in size.

(*b*) The Nuclear Bead

The nucleus is the major requirement in pearl culture operations. The basic new material to produce nuclei is a species of bivalve molluscs of family unioidae.

Steps Involved

Clearing the shell

⬇

Drying the shell

⬇

Making a fine powder

⬇

Mixing with fixative (silica gel etc)

⬇

The desired shape is obtained with the help of icons.

Methods of Implantation

- ☆ The CIFA has involved various surgical procedures for obtaining different freshwater pearl products :
  1. Mantle cavity insertion method
  2. Mantle tissue implantation procedure
  3. Gonadal implantation surgery.

Mantle Cavity Insertion Method

☆ In the mantle cavity insertion method the appropriate nuclei up to 1 cm are placed in the mantle cavity of pearl mussel of size 8-10 cm in shell length and implanted mussels reared for a period of 12 month in pond culture environment.

Products Obtained

☆ The products generally are shell attached, half round or design pearls depending upon the shape of the nuclei implanted.

Mantle Tissue Implantation Procedure

☆ In mantle tissue implantation procedure a donor mantle graft soaked in diluted cosin of 2-3cm^2 alone (Mantle tissue non-nucleated insertion-free irregular pearls) or with a small nucleus (below 3mm die, mantle tissue nucleated–small round pearls) is placed in the space between the outer and inner epithelial layers of left and right mantle lobes of recipient mussels.

☆ The implanted mussels transferred to the pond environment for culture for a period of twelve months or more.

Products Obtained

☆ The products are unattached irregular to oval graft pearls on small round nucleated pearls.

Gonadal Implantation Surgery

☆ In gonadal implanted surgery, a donor mantle graft (2 mm^2) along with a nucleus (3-6 mm dia) has to be implanted in the gonad of the recipient mussels. The gonad implanted mussels require post operative care with antibiotic support under controlled conditions for 7-10 days to minimize the graft and nucleus rejection rate. The pond culture phase of the gonad implanted mussels varies from 6-12 months depending on the size and number of nuclei implanted.

Products Obtained

- ☆ The products are unattached, and slightly larger, round culture pearls.

Post Operative Management or Rearing of Implanted Mussels

The following criteria have been found to be very useful in production of quality pearls with attractive colour.

- ☆ Rearing pond: The pond should preferably be rectangular in shape, perennial in nature, measuring about 0.1-0.2 ha. with a water depth of 2-2.5 m depth and the visibility of 30-60 cm. The water colour is to be yellowish green indicating diatoms and sufficient green algal content.
- ☆ The rearing tank should be free of macro-vegetation.
- ☆ To begin with the pond should be prepared with fish toxicant like mahua oil cake at the rate of 2500kg/ha/m to get rid of unwanted organisms and harmful carnivores.
- ☆ Ca and Mg salts are the most important factor in pearl production, with Ca of optimum level 16 ppm closely followed by Mg. The role of micronutrients is no less important in pearl production.

☆ The ranges of physico-chemical parameters of water of rearing ponds are as follows:

Temperature	- 15-28°C
Transparency	- 30-40cm
pH	- 7-8
Water depth	- 2m
Total alkalinity	- 70-150 ppm
Hardness	- 100-200 ppm
Phosphate	- 0.36-0.5 ppm
Nitrate-Nitrogen	- 0.48-0.62 ppm
Ammonia (Unionized)	- 0.022-0.055 ppm

☆ The ranges of physico-chemical parameters of rearing pond soil are as given:

pH	- 7-7.5
Organic carbon	- 1-1.5 per cent
Available N_2	- 25mg-500mg/100gm
Available Phosphate-phosphorus	- 20-25 mg/100gm
Free $CaCO_3$	- 1.5-3 per cent

Rearing Methods

☆ Better results are obtained when implanted mussels are kept in suspension underneath the water.

☆ Implanted mussels are first kept under 'Intensive rearing' in nylon bags for 10-12 days.

☆ Each mussel is kept in one nylon bag with their open side upwards.

- ☆ These bags are tied to the bamboo poles fixed in a pond.
- ☆ Implanted mussels are transferred from nylon bags to cages hanging with bamboo poles and reared further in manured ponds for 1-4 years.
- ☆ Mussel feeds on unicellular algae and particulate organic matter.
- ☆ For growth of algae, the pond should be manured with cow dung @ 10000 kg/ha/yr, Urea @ 100 kg/ha/yr and single super phosphate @ 100 kg/ha/yr.
- ☆ Stocking density of 0.5-1 lakh mussel/ha can be followed depending upon carrying capacity of pond.

Harvesting

- ☆ December to February is the ideal period for harvesting of pearls.
- ☆ Harvest is done after a year or even after 3-4 years.

Procedure for Implantation

- ☆ The collected mussels should be thoroughly cleaned and kept immersed in a tray ventrally upward to facilitate opening of the shell-values.
- ☆ A piece of cloth smeared with a pinch of menthol shall also be kept submerged in the water of the tray containing mussel.
- ☆ A portion of the mantle epithelium of pallial region (posterior to middle) is cut out and made into small pieces with the help of a sharp knife on a glass slide to form graft tissue for implantation. They

are then cleaned by a sponge, stretched and trimmed properly to size of 2-4 mm.

- ☆ The cut pieces are soaked in Neosporin/Cosin/ physiological saline for keeping them potent and free from bacterial infection and ready for implantation. The cut pieces (*i.e.* graft tissues) should be covered with watch glass to prevent dehydration and should be used up for the purpose of insertion soon, but not within an hour.
- ☆ The size of tissue (cut pieces) should be between 4-6 mm in the case of non-nucleated method of implantation. But in the case of nucleated method, the size of the graft tissue should be 1/3rd of the beads to be implanted or approximately of 2-4 mm.
- ☆ The shell-valves of the recipient mussel (mother mussel) have to be opened by means of a shell valve opener and wooden peg is to be introduced to keep the valves open (within 1.5cm) for the purpose of insertion.
- ☆ The mussel is than to be attached with firmly to the mussel-holder.
- ☆ Small slit is to be made either inside the mantle or inside the gonad with great care by means of sharp incisions knife.
- ☆ The bead, soaked in mercurochrome solution, is then to be placed in the slit by means of a nucleus carrier.
- ☆ At this position, the graft tissue is to be so placed on the implanted bead that the dorsal portion of the tissue rests on the bead in the mouth of the wound.

- ☆ The slit is now to be closed by the blunt end of the middle in such a way that the original position is restored.
- ☆ The blunt end of middle is to be dipped into mercurochrome solution (antiseptic) before use.
- ☆ The mussel is then to be dislodged from the holder and the peg is pulled to allow the shell valves to close.
- ☆ The implanted mussel is to be placed inside a hapa containing water after thorough clean and left overnight for conditioning.
- ☆ On the fall day the operated live mussel are to be segregated from the dead ones if any, after close moderation and then released back into the culture pond or cage. Care should be taken the mussel do not overlap.
- ☆ After a month or so, the mussels are to be taken out from the pond to study whether the inserted beads and tissues are in a proper place or not.
- ☆ Formation of pearl through non-nucleated method takes about 1.2-2 years while by nucleated method it would be 1 year or so.

Harvested Pearls are Graded According to their Size and Lusture

Minute granular	:	>2.6 mm
Fine	:	2.6 to 4.9 mm
Small	:	5.0 to 6.8 mm
Medium	:	6.9 to 8.2 mm
Large	:	>8.2 mm

Chapter 19

Organic Farming

In aquaculture, sustainability means not only the development in terms of output but also the socio-economic and ecological parameters. As the availability of land and quality of land are decreased day by day by application of in-organic fertilizer pesticides, herbicides and antibiotics etc. The effect of prolonged and over usage of chemicals on soil results in human hazards and pollution of the environment. Hence the expectation that organic farming by reverting to the use of manure, green manures, urban and rural wastes etc., can enhance sustainability to aquaculture with eco-friendliness. The globalization of trade under world trade organization and opening of market to international players has thrown new challenges to Indian farmers. Till today all our effects have been aimed at one goal increase the production to maximum.

Organic farming is defined as a production system, which avoids or largely excludes the use of synthetically compared fertilizer; pesticides, growth regulators and livestock feed additives. To the maximum extent feasible organic farming system depends on crop rotation, animal manures, legumes and green manures.

Types of Organic Farming

Pure Organic Farming

This farming can be achieved by use of organic manures and biopesticides and complete avoidance of inorganic as well as pesticides, antibiotics etc.

Integrated Farming System

This is done by the low input organic farming in which the available resource are effectively utilized and recycled. This involves cropping along with poultry fishpond, mushroom, goat rearing etc. apart from crop components.

Animal Based Manures

These include farmyard manure, biogas slurry, poultry manure etc. Even though organic manure contain low NPK as compared to inorganic fertilizers, it is superior by virtue of supply of micronutrients, improves microbial activity and by production of organic acids, which binds soil particles, improves physical condition of soil. They hasten the growth and development by supplying growth regulators. Organic manure act as slow releasing fertilizes because of slow release of ammonia and slow conversion of nitrates; the leaching losses of N are low in the presence of organic manure. Experiments conducted at Agriculture Research Station, Siruguppa revealed that among different organic measures, poultry manure was the best. One ton of poultry manure was equivalent to 7 tons of farmyard manure. Cow dung is the main source of FYM, which contains 5kg N, 1.5kg P and 5 kg per tonne, and it also improves the soil health.

Vermicompost

It provides nitrogen, phosphorus, potassium, calcium and magnesium and micronutrients such as Fe, Mn, Zn, Cu, etc. It also contains growth-promoting substances such as NAA, Cytokines, Gibberlins etc.

Plant Based Manures

Green manures (*Sesbania* sp.; sun hemp), green leaf manure (*Pongamia, neem*), oil cakes (ground nut, sunflower) and crop residues (sugarcane trash, cotton stover), contain low NPK and high C : N ratio, improve the physical, chemical and biological activities in soil, increase the water holding capacity.

Plants extracts like neem, pongamia, tulsi, sorghum, marigold and bougainvillea have pesticidal effect and were found to be effective against number of pests and diseases. Neem oil has been recommended to mix with other pesticides to improve its efficiency.

Bio-herbicides: The term bio-herbicides refer to microbial plant pathogens, which are adopted as spray to kill or suppress the growth of weeds. They have the following advantages:

- ☆ A high degree of specificity for the target weed.
- ☆ Safe to non-target and beneficial plants.
- ☆ Absence of weed resistance development.
- ☆ Absence of residues build up in the soil or biosphere.

There are several biological agents like insects, fungi and bacteria etc., which are effective against number of weeds and are proving effective against water hyacinth and a beetle.

Bio-fertilizer: they convert atmosphere N into available form and solubilize soil phosphorous such as Blue green algae or cyanobacteria. *Azolla* is an aquatic fern widely found both in temperate and tropical natural conditions. *Azolla* fixes atmospheric nitrogen in symbiotic association with Anabaena algae. This photosynthesis is to fix atmospheric nitrogen up to 100–150 kg/ha annually. It is more widely used now in paddy cultivation for better crop production. Azolla could be used either as green manure before transplanting or as a dual crop after transplanting of rice. The growth and efficiency of Azolla in general is high in hilly or semi-hilly humid areas or in coastal areas than in regions with high temperature. Application of algal bio-fertilizers is also useful for the reclamation of marginal solid such as saline-alkaline and calcareous soils.

Benefits of Organic Farming

- ☆ Conservation of Natural Resources.
- ☆ Prevention of damage to environment.
- ☆ Production of clean foods.
- ☆ Reduction of entry of toxicants in the food chain.
- ☆ Prospects of export of organically produced foods.
- ☆ Sustaining production system dependent largely on on-farm resources.
- ☆ Prevent soil erosion and degradation.
- ☆ Reduction in pollution of water bodies.

Standards for Organic Aquaculture

Germany has framed certain standards for organic aquaculture registered with Naturland association. It has been adopted nationally and internationally and

International Federation of Organic Aquaculture Movements also confirms that the standards set by Naturland fulfill the requirements of an international organization. These standards are classified into two categories *viz.*, general principles of farm management and Miscellaneous.

General Principles of Organic Aquaculture

Selection of Farm, Site Location with Surrounding Eco-systems

Site should be free from effluents caused by some other sources which may be threatened to negative impact. Farms should be protected by adopting suitable preventive measures by avoiding predatory birds and other animals.

Species and Origin of Stock

In organic farming, the native species can be selected for culture and the indigenous species should be avoided. Other wise the seeds has to be purchased from conventional suppliers with following conditions *viz.*, organisms should not be genetically obtained by means of ploidy induction.

Breeding

Breeding should be taken place under natural conditions. The use of hormones, even from the same species is not allowed.

Design of Holding System, Water Quality and Stocking Density

This conditions will ensure the specific behavioural pattern which include movement, feeding and reproduction habit. The water quality must fulfill the natural requirements of the species. Materials and substances used in the constructions and management must not cause any injuries effect on the organisms or the environment.

Health and Hygiene

This can be achieved by adopting natural curative and preventive measures. The stock can be regularly monitored and assessed for the health status.

Oxygen Supply

This can be provided from the natural, physical conditions of the water body.

Organic Fertilizing

This can be done by applying organic manures such as cowdung, poulty manure etc.

Table 13

Finfish	*Shellfish*
☆ *Lates calcarifer*	☆ *Scylla* sp.
☆ *Grouper* sp.	☆ *Panaeus indicus*
☆ *Mugil* sp.	☆ *Panaeus monodon*
☆ *Puntioplites/Puntius* sp.	☆ *Haliotis* spp.
☆ *Ophiocephalus micropeltes*	☆ *Ostrea rivularia*
☆ *Pangasius bocourti*	
☆ *Oreochromis niloticus*	
☆ *Aristichthys nobilis*	
☆ *Ctenopharyngodon idella*	

Feeding

Feed should be free from the GMO incorporation. Natural pigments are permitted and addition of synthetic hormones or substances must be avoided.

Transport

Live fishes must be provided with adequate oxygen during their transportation. The fishes should be starved before their transportation. A transport density of one kg of fish to eight litres of water shall not exceed. The water must be changed after 6 hours of transport taking care to maintain the same temperature. Transport duration of 10 hours shall not be exceeded.

Chapter 20

Pen Culture

Raising of fish in confinements of water areas in large water sheets is a specialized aquaculture technique which is gaining importance world over for intensive fish production. This aquaculture system consists of a netting or screening, which allows free circulation of water for growing of fry, fingerlings, yearlings or further advanced stages of finfishes or shellfishes to a large marketable size. This system is in the form of cages, pens or fixed enclosures.

A pen is generally set up along margin of capture fishery water bodies. Farming of aquatic organisms is practiced in this particular type of rearing facility with the sides enclosed by bamboo fencing, wooden poles or netting. The bottom of the structure, however, will continue to have the natural substratum.

The origin of pen culture is recent. According to Alferex (1977), pen culture was originated in the inland sea area of Japan in the early 1920s. It was adopted by the People's Republic of China in the early 1950s for rearing carps in freshwater lakes. It was introduced in Laguna Bay and the San Pablo Lakes in the Philippines by the Bureau of Fisheries and Aquatic Resources (BFAR) and the Laguna Lake Development Authority (LLDA) of that country between 1968 and 1970 in order to rear milkfish, *Chanos chanos* (PCARRD, 1981). This practice spread to other Southeast Asian countries like Indonesia later on. Presently, commercial pen culture of fish is in vogue in the Philippines, Indonesia and China.

Very little work has been done regarding pen culture in India. However, traditional trapping and extensive culture of tiger prawn (*Penaeus monodon*), milkfish (*Chanos chanos*), Pearlspot (*Etroplus suratensis*), mullet (*Mugil* sp.), bhetki (*Lates calcarifer*) and threadfin (*Polynemus* sp.) are done in improvised pens and enclosures in Kerala and in the shallow areas of Chilka Lake in Orissa. In situ carp seed rearing in pens was attempted in Bhavanisagar and Tungabhadra reservoirs in 1978 for rearing fingerlings.

Scope of Pen Culture

The country's requirement of fish seed is very high and there is an urgent need to diversify the fry rearing systems from the traditional farm pond practice in order to overcome the present seed supply crisis impeding aquaculture expansion. India has vast habitat resources for aquaculture in terms (*a*) ponds and tanks (2.36 million ha), (*b*) oxbow lake and derelict water bodies (1.30 million ha), (*c*) reservoirs (2.05 million ha), (*d*) rivers–45,000 km, canals (1,26,334 km), (*e*) estuaries (2.7 million ha) and brackish water (1.42 million ha) (Ayyappan and Venkateshwarlu, 2002). These resources have been hitherto under-exploited and need to be harnessed for augmenting fish production by taking up different aquaculture practices, including pen culture of suitable fishes. Cyprinids, perches, murrels, catfishes etc., can be easily cultivated in pens.

Types of Pens

There are three types of pens. These are:

1. Completely isolated enclosures surrounded by a net structure in the middle of a bay with no foreshore,
2. Shore or lake enclosure with a portion of the foreshore extending into deep water surrounded by a net structure, and
3. Bay or lake enclosure with an embankment provided with sluices or net structure only at the entrance.

Site Selection

Before installation of a pen, proper site selection is the most essential part for successful enclosure aquaculture.

Figure 21: Pen

Sheltered, weed-free shallow water bodies are the ideal locations for installing pens. The other criteria of area needed for pen culture are:

1. Pollution free zone
2. Moderate wind action or wave action
3. Good soil and water quality
4. Water depth (0.5-2.0m)
5. Adequate water circulation
6. Sufficient availability of quality seed
7. Market accessibility and
8. Ownership/leasing.

Water level fluctuation is the most important consideration in pen culture operation in the lakes, especially the man-made ones. A scrutiny of contour map and the monthly fluctuation patterns of reservoir levels will enable location of the suitable sites, which would retain sufficient water for the required period of time. Sites, which dry off during summer will be ideal as it will be easier to erect pens on the dry land, to be inundated later as the water level increases.

Pen Materials and Construction

The materials required for pen construction are:

1. Fencing screen of suitable material with following features:
   (*a*) Mesh size of the screen material must be small enough to retain fish fry or fingerlings
   (*b*) Sturdy enough to withstand the stretching tension, current, wind and wave action

(c) Rot, corrosion and weather resistant

(d) Resistant to cuts by crabs and other animals

(e) Relatively cheap and easily available and

(f) Easy to handle and transport.

2. Bamboo or wooden poles to support and hold the pen screen in position against water current, wind and wave action.
3. Foot and head ropes of 3 to 5 mm dia of durable quality.
4. Tying or lacing twine of about 1 mm thickness, and
5. Bricks for keeping the bottom line well inside the soil.

Metal and metal coated with HDPE screens are often used for pens in western countries and Japan, which are although highly durable, are cost prohibitive in developing countries.

The method of erection of pens can be of two types *viz.*, on wet ground where already water exists and, on dry ground, which gets dried up during summer months but during monsoon submersion up to 1.2 m is expected. Depending upon the area and contour of water body the shape and sizes of pen structures are variable. The enclosure screen in Southeast Asian countries is usually made of split bamboo mats, supported by vertical stalking poles fixed at 1 to 2 m intervals. The mats are interspersed through the poles and secured by coir ropes or nylon twines. A trench is excavated and the lower part of the screen is partially buried inside and filled, to prevent entry or escape of fishes from the bottom. A nylon netting or monofilament screen of suitable mesh size is provided inside for additional safety of the stock. However, monofilament screens reinforced with bamboo framework only are used for nursery pens. They should have a free board of about 1.0 m above the normal water level to prevent entry or escape of fishes by jumping and as a precaution against water level fluctuations.

Cultivable Species in Pens

The market demand and the availability of seed greatly influence the selection of candidate species in pen culture. However, the main desirable characteristics of the candidate species for enclosure aquaculture are faster growth rate, high survivability, capacity to withstand overcrowding, acceptance of artificial feeds, high feed conversion ratio and resistance to diseases. Cultivable finfishes and shellfishes of freshwater, brackish water and marine water can be cultured in pens.

The Indian and Chinese carps (*Catla catla, Labeo rohita, Labeo calbasu, Cirrhinus mrigala, Cyprinus carpio, Hypophthalmichthys molitrix* and *Ctenphyaryngodon idella*), air breathing catfishes (*Clarias batrachus* and *Heteropneustes fossilis*), tilapia (*Oreochromis mossambicus*), murrels (*Channa striatus* and *Channa marulius*), freshwater prawns (*Macrobrachium rosenbergii* and *Macrobrachium malcolmsonii*), Milk fish (*Chanos chanos*), pearlspot (*Etroplus suratensis*), mullet (*Mugil* sp.), bhetki (*Lates calcarifer*), threadfins (*Polynemus* sp.), tiger shrimp (*Penaeus monodon*) are suitable candidate species with potential for being reared in pens. Mono or polyculture of these species can be practiced.

Routine Management of Pens

The following aspects are important in the routine management of pens:

1. Feeding the fish with a cost effective, nutritionally balanced feed with an appropriate feeding schedule
2. Regular monitoring of water quality parameters
3. Maintenance and periodic cleaning of net materials from fouling organisms
4. Regular monitoring of growth and diseases by sampling
5. Continuous surveillance to check poaching, and
6. Regular investigation to prevent escapement of cultured organisms through any damaged area of pen.

Pen Culture: Advantages and Disadvantages

Advantages:

The following are the advantages:

1. High production per unit area can be obtained because of dense stocking and intensive feeding
2. There is possibility of making utmost use of all available water bodies
3. Predatory and non-predatory fishes can be cultured in different pens of the same water body
4. Feeding and monitoring are easier here
5. Metabolites and wastes from the pens are washed out by the free exchange of water or wave action, thus facilitating higher stocking rates

6. Chances of pollution are very less here
7. Pens are fairly simple to construct using local, unskilled labour
8 It provides gainful employment and generates income.
9. The left over feeds and faecal matter enrich the main water body, thereby increasing the productivity.

Disadvantages

The disadvantages are:

1. Pens are generally fouled by sessile organisms and therefore, regular cleaning is required, which is a labour intensive job
2. There is always danger of fish escaping through openings in the netting caused by other animals such as crabs
3. The material used mostly for pen construction is bamboo, which is prone to decay in water. So there is a need to replace the pen with a new one after each operation cycle
4. There may be problem regarding ownership of the water bodies and securing rights for setting up pens
5. In the public water bodies, the farmers may be helpless to stop or control any pollution that occurs
6. Pens are susceptible to storm damage compared to ponds, tanks or raceways and
7. Risk of theft and vandalism.

In India, reports on fish culture in pens either in freshwater or brackish water are very few (Lalmohan, 1983; Laal *et al.*, 1990; Routray and Routray, 1998). All these studies are restricted only to production of fry and fingerlings. The Central Inland Fisheries Research Institute (CICFRI), Barrackpore, has standardized the methods of culture of freshwater prawn, *Macrobrachium rosenbergii* in pens (Sugunan and Sinha, 2001). Captive rearing system holds great potential, where land for construction of ponds is either expensive or is not available and suitable watersheds are not being fully utilized. Considering the availability of vast water bodies in the form of reservoirs, lakes, beels, flood plains and wet lands in the country, the technology of pen culture would be an effective tool for the development of these large water bodies. There is also need for development and standardization of pen culture practices in relation to different water bodies. Dissemination of the pen culture technology by fisheries scientific and extension personnel sector of the Central and State Governments will go a long way in increasing the current fish production level through popularization of this technology.

Chapter 21

Cage Culture

India has vast marine as well as fresh water resources, having immense potential to be utilized for aquaculture too. Marine resources of India cover sea area along a coastline of 8185 km. In addition there are other related resources comprising backwaters, brackish water impoundments, lagoons and estuaries, besides mangroves. In all, 1.42 million ha of these waters, a substantial part of which is beyond coastal regulation zone (CRZ) could be developed for undertaking sustainable and environment-friendly coastal aquaculture. Similarly, inland waters in the form of 14 large, 44 medium and many small rivers, 1.2 lakh ha of irrigation canals, 2.90 mill ha of lakes and reservoirs, 1.30 mill ha of beels and derelict water bodies and 2.85 mill ha of ponds and tanks can offer a sizeable area for taking up aquaculture activities (Pathak, 1999). In addition to existing culture practices, reservoirs, beels, irrigation canals etc., could be utilized for fish cage culture.

Origin

Development of cage culture practice seems to have developed independently in a number of South-East Asian countries, *e.g.*, floating cages have been in use since the end of the 19th century in the Great Lake region of Kampuchea (Pantulu, 1979). This practice spread to neighbouring Vietnam and Indonesia in early 20th century (Reksalegore, 1979). Modern cage fish farming probably first began in Japan in the early 1950s by the efforts of Prof. Harada, Director of Fisheries Laboratory at Kinki University, Japan, who started experimenting with yellow tail cage culture in 1954. Sampath and Menon (1975) succeeded in rearing *P.monodon,* in small (4.5 × 2 × 1 m) fixed cages in estuaries near Kochi. Since then penaeid species have been successfully cultured in cages in other parts of India (Maheswari, 1984).

Operation Principles

One of the main advantages of culturing fish in cages is that they do not compete for bottom space as they do when cultured in ponds. Generally, the fish production in one ha of cages is approximately equivalent to that in 12-50 ha of ponds, depending on water flow etc. The reason is that fish cages allow for a free exchange of water to maintain higher dissolved oxygen levels than in a pond, and also allow fish wastes and leftover feed to exit the cage without accumulating. Therefore, oxygen depletion or deterioration of water quality, even at high stocking densities does not stress caged fish. Furthermore, the fish assimilate feed more efficiently, resulting in higher production per unit of input. In general, the moving out of waste materials from the cages to the surrounding water body is not a problem so long as the ability of the receiving water to assimilate the waste is not exceeded.

Cage Design

Cage design depends upon various factors like species to be cultured, prevailing environmental conditions, method of culture (extensive/semi-intensive/intensive), costs and availability of materials as well as local skills. Two main components of cage design are: a) cage structure and b) cage size.

Cage Structure

Cages are composed of a frame, netting, floats, sinkers, and anchors. The frame supports the cage netting, maintains the shape of the cage, and acts as a float or operational platform. Traditional cages are fabricated using existing skills from locally available material like grasses or wood. Bamboo is a much more suitable material being strong, cheap, widely available and easy to use. But these materials decompose at a faster rate in tropical waters. They can also be damaged by the caged fish and predators. Modern cages utilize materials such as nylon, plastic, polyethylene and steel mesh which are, though more

Figure 22: Cage

expensive, have a much longer life span and permit better water exchange.

Cage Size

Cage size is ultimately based on a compromise between cost of fabrication, operational costs and per unit area production, and is aimed at maximizing economics and efficiency of production efficiency. It depends on the following factors:

Area

The larger the cage area, the lower the production cost. Here, the area of net material required is 30 per cent higher, when four small cages are constructed rather than one large cage with the same total surface area. A possible disadvantage with larger cages, on the other hand, is reduction in water exchange with the surrounding environment and when natural food sources are utilized, the productivity of the larger cage is lower.

Cage Placement Depth

If natural food sources are primarily used to feed fish, cage placement at certain depth should be based on the vertical distribution of plankton in the water. The recommended cage depth is 2-3m, for the reason that in reservoirs, over 60 per cent of primary productivity is found between the depths of 0-3m. The cage bottom should be at least 0.5m above the lake or reservoir bottom to allow free exchange of water and for discard of fish wastes and leftover feeds.

Mesh Size

Cage mesh size should be selected in such a way that stocked fish cannot escape, and a better water exchange is ensured.

Shape

Round cages are more efficiently used by the schooling fishes such as the milkfish, which tend to swim in circles when enclosed and are lower in cost. However, these cages are more difficult to manufacture and maintain. Square shaped cages are least efficiently used by fish. Rectangular cages are recommended because of their simple structure, ease of installation, wide adaptability and low cost.

Depth

The influences of surface area and depth of cages on the health and production of fish have been little studied. Cages of less than 1.5m depth have been shown to retard growth and change body shape of common carp and tilapia (Maruyama and Ishida, 1977), while depths greater than 10-12 m are probably poorly utilized by most fish species except during period of adverse surface temperature or water quality conditions. Cage bag depths of between 0.9 and 1.6 m are recommended as they give the fish sufficient shelter from surface effects, also providing an adequate water volume for exercise and feeding (Kerr *et al.*, 1980).

Cage Types

There are four basic types of cages as shown in Table 14.

Site Selection

The correct choice of site for cage fish farming is vitally important, since it can greatly influence economic viability as determined by estimated capital outlay, running costs, rate of production and mortality factors. The criteria for selection of suitable sites for cage culture are shown in Table 15.

Table 14: Types of Cages and their Characteristics

Type of Cage	*Characteristics*
Fixed Cage	☆ Consists of net bag supported by posts driven and fixed into the bottom of a lake or river
	☆ Limited in size and shape
	☆ Restricted to sheltered shallow sites with suitable substrates
Floating Cage	☆ Most widely used cage type
	☆ A frame supports the cage bag
	☆ Floats are attached to the frame for buoyancy
	☆ Can be designed in a variety of shapes and sizes according to choice of the culturist
Submersible Cage	☆ Relies on the frame or rigging to maintain shape
	☆ Position in the water column can be adjusted to take advantage of prevailing environmental condition
Submerged Cage	☆ Structure is similar to floating type but does not have floats
	☆ Anchored to the substrate by stones or posts

Culture Types

Cage culture can be classified on the basis of feed inputs based on chosen adoption of extensive, semi-intensive and intensive culture systems. Extensive cage culture is restricted to freshwaters and may be practiced in two types of environment *viz.*, highly productive lakes and reservoirs, and water bodies that receive considerable quantities of sewage or domestic wastes. Tropical water bodies with high nutrient loading offer the best opportunities for extensive cage culture. Commercial extensive cage culture is at present widely practiced only in the Philippines and in China. Planktivores, detrivores and omnivores, such as many of

Table 15: Site Selection Criteria for Cage Culture

No.	*Environmental Criteria for Cultured organisms*	*Environmental Criteria for Cages*	*Site Facilities Required and Management*
a)	**Water quality:** Pollution free water with optimum physico-chemical parameters according to species requirement	**Weather:** Sites with violent storms and extreme cold conditions should be avoided	1. Site ownership 2. Licensing 3. Care of surrounding environment 4. Navigational problems 5. Availability of freshwater 6. 3-phase power supply 7. Telephone services 8. Transportation through road/railway 9. Security of cages
b)	**Phytoplankton blooms:** Toxic blooms of cyanobacteria and dinoflagellates in fresh and marine waters respectively cause fish kills and also cause DO depletion. Such sites should be avoided.	**Water Currents:** 1. Ensure good water exchange and food supply 2. Provide addition dynamic loading to the cage structure due to excessive currents 3. Sites with water current <100 cm/s are preferable to check the feed loss	

Contd...

Table 15–*Contd...*

No.	*Environmental Criteria for Cultured organisms*	*Environmental Criteria for Cages*	*Site Facilities Required and Management*
c)	**Disease:** Sites where disease agents are present in the wild fish, intermediary hosts or environment should be avoided but introduction of diseased stock to the farm or attraction of birds may still cause disease.	**Depth:** Cages should be installed in sufficient depth to maximize the exchange of water and to kept the bottom of the cage clear of the substrate.	
d)	**Water exchange:** This is essential at intensive culture site to minimize build up of wastes, where as at semi-intensive and extensive culture site, poor water exchange is preferred to utilize primary productivity.	**Substrate:** Muddy to sandy ground allows use of fixed cages in freshwaters. Rocky substrate is good for marine sites since it indicates good current with reduced risk of waste build up.	
e)	**Fouling:** Due to this, mesh size decreases, DO supply reduces due to poor water exchange and waste accumulates. Rate of fouling increases with increase in salinity.		

the carps and tilapias, may be suitable candidates for extensive culture. In the Philippines, production of tilapia by means of extensive cage culture is claimed to be very high ~up to 1.90 kg/cu m/month.

Semi-intensive cage culture is done to rear species that feed low in the food chain such as tilapia and bighead, silver and common carp. These fishes are fed an almost infinite variety of materials including rice bran, brewery and domestic wastes. Most types of freshwater sites have proved to be suitable for semi-intensive cage culture, although precautions may have to be taken to avoid excessive feed losses. However, semi-intensive culture is not practiced in marine environments.

Intensive cage culture is largely restricted to the rearing of high value carnivorous species. In freshwater, salmonids, channel catfish and increasingly common carp are cultured, while in the marine environment, Atlantic salmon, yellowtail and groupers are the main intensively farmed species. Intensive culture is not recommended in lotic sites, where feed losses can be excessive.

Carrying Capacity

The maximum capacity of a site to sustain the production at that level is called as the carrying capacity of the site. This should be a major consideration for the selection of a suitable site for cage culture, since it is directly related to fish production. Intensive cage culture results in the production of wastes, which can stimulate productivity and alter the abiotic and biotic characteristics of the water body concerned. Content of phosphorus in freshwater sites and Nitrogen in marine sites are considered as the limiting nutrients, which control phytoplankton abundance.

Increase in Phosphorus level through leaching from feed results in plankton blooms and hence, it indirectly affects water quality and thereby altering carrying capacity. Since marine coastal sites generally have much higher flushing rates compared to inland water sites, the effect of dissolved Nitrogen on phytoplankton production may be limited.

Carrying capacity of inland waters for semi-intensive culture depends on (i) the productivity of the water body, (ii) the amount of natural food available, and (iii) the quantity and quality of supplementary food used.

In the case of extensive cage culture operation, carrying capacity is almost dependent upon phytoplankton production externally. According to Beveridge (1984b), yields from extensive cage culture are between 1 per cent to 3 per cent primary production.

Stocking Density

Stocking density of fish depends on various factors like species to be cultured, size of fish, cage volume, water quality, type of culture practice (intensive/semi-intensive/extensive) and level of production to be obtained. In the case of shrimps and prawns, the stocking density generally depends on the surface area of the cage instead of cage volume because of their benthic habitat.

Feeding

At most of the intensive and semi-intensive cage farms, fish are fed throughout the year, except during periods of adverse weather. Most cage farmers feed their fish manually, using supplementary feeds, trash fish or moist feeds, since this is cheaper and easier to do so, at least on a small scale. Feeding with frozen trash fish by thawing is advantageous since bacterial infections and water pollution are reduced (Taniguchi, 1983). Dry supplementary feeds and moist pelleted feeds compounded from trash fish/fish meal with binder are also fed to caged fish in a similar manner. By manual feeding, farmers can assess how hungry their fish are and consequently can adjust the amount of feed they use but this is very labour intensive.

In recent years, mechanical feeders have been put into use at most large intensive cage farms for saving expenditure on labour. There are two types of mechanical feeders–Demand feeders and Automatic feeders. Demand feeders work on the principle of the demand of fish for feeding while automatic feeders deliver measured ratio at fixed intervals, which are preset by the farmer using a timing device.

Routine Management

Monitoring Water Quality

Daily measurements, especially of dissolved Oxygen and temperature should be recorded, preferably at dawn and mid-day and at slack tide in the case of marine cages when the highest and the lowest levels are found respectively. The readings both inside and outside cages and at cage surface and cage bottom should be taken. Data on nitrogen (ammonia, nitrate, nitrite) and dissolved phosphorus levels, pH, secchi disc and chlorophyll levels are also of importance.

Sampling

Samples of fish should be taken at regular intervals, their length and weight is to be noted, so that the growth of stocks can be monitored. Survival and mortality are to be recorded at each time, which will help in determining stocking density and feeding rate and also the time of harvesting. As the fish grow, periodical thinning is also necessary to maintain optimum growth conditions and for minimizing disease risks. Regular monitoring of fish stocks from a disease point of view is also essential.

Harvesting

Harvesting of fish is either done continually or in batches, depending on how the production cycle is geared. Prior to the harvesting of a batch, the fish should be starved for one or two days, in order to firm up the flesh and to evacuate the gut contents, since a gut full of food and bacteria can accelerate deterioration and increase the risk of carcass contamination during the gutting process. Harvesting is done by pulling the net up until the fish are concentrated in a small volume of water and netted out using dip nets.

Maintenance of Cages

Cage moorings have some definite life span. Hence they must be checked at intervals for signs of damage and repaired or replaced when necessary, as otherwise they can put the cages and fish stock therein at risk. At most sites, particularly in the marine environment, nets become fouled and have to be changed and cleaned. Chemical and mechanical methods are applied to clean fouled nets. Mostly the nets are laid out to dry for a few days, thus making them easier to clean with the help of hard bristle brushes or high-pressure hoses. Chemical methods include soaking of nets for 2-3 days in a 3 per cent Formic Acid with 9 per cent Copper Sulphate solution, or for 3 hours in Sodium Hypo chlorite solution (Sutterlin and Merrill, 1978; Moller, 1979).

Environmental Impact

The introduction of cage to a water body will have an impact on the environment, which can lead to conflict, since inland waters are increasingly under pressure from other users for a wide variety of purposes. Cage structures affect a water body in three principal ways: 1) They take up space, thus potentially competing with other users: 2) They alter flow regimes which govern the transport of oxygen, sediment, plankton and fish larvae; and, 3) They have an impact on the aesthetic qualities of the site. Irrespective of the method of rearing, cage culture can introduce or disrupt disease and parasite cycles, change the aquatic flora and fauna and alter the behaviour and distribution of local fishes. Associated with intensive cage culture is the release of uneaten food and faecal matter into the environment, stimulating primary production but adversely affecting water quality.

Cage culture: Advantages and Disadvantages

Advantages

- ☆ It is easier to control competitors and predators in cages. Also, cages are convenient for the prevention and treatment of diseases.
- ☆ Harvesting is easier than in ponds and is also cost effective.
- ☆ Fish cages are portable and easily shifted from one place to another when necessary and they require less capital investment for construction, have lower production costs, and produce higher economic returns compared to ponds.

- ☆ Majority of cages are fairly simple to construct and can be put together in a day of so using local, unskilled labour. They are also easily managed.
- ☆ Numerous studies have confirmed that cage-reared fish are often superior to fish reared in other systems and even wild fish, in terms of condition factor, appearance and taste (Holt *et al.*, 1978; NORDA, 1984).
- ☆ Cages make use of existing water bodies. This can provide an opportunity for non land-owning sectors of the fisher community to utilize the area for fish farming, thus greatly increasing their income.

Disadvantages

- ☆ The production of fish in cages is dependent on the ecological conditions of the water.
- ☆ Cages become fouled with sessile organisms, and to maintain an unrestricted exchange of water, regular cleaning is required, which is labour-intensive task.
- ☆ Unlike ponds, food, which is uneaten by the fish, cannot be recycled, and is lost from the culture system.
- ☆ There is always the danger of fish escaping, as animals like crabs may damage the cage netting.
- ☆ The major disadvantage of cage fish farm is that they are often sited in public or multi-user water bodies and farmers may be powerless to stop or control any pollution that occurs.

- ☆ Cages are more susceptible to storm damage than ponds, tanks or raceways.
- ☆ Cages are highly vulnerable to poaching and vandalism.

At present, contribution of cage culture to world farmed fish, shellfish and crustacean production is only around 4 per cent. Though several efforts have been made so far in India to launch commercial cage culture in both marine as well as freshwater areas, the activity remains restricted to an experimental level only and it could not create the needed impact among fish farmers. Some of the reasons for this may be enlisted as lack of dissemination of standard cage culture technology to the potential fish farmers, short-term lease of culture sites, short life span of cages and high depreciation of cage material and, most importantly, problems of poaching and vandalism. Due to these problems, cage culture has remained a non-bankable proposition as financial institutions are not coming to provide financial support to cage culture systems. Considering these points, dissemination of cage culture technology deserves serious attention from scientific and extension community of the fisheries sector as well as central and state governments.

Chapter 22

Sewage-Fed Fish Culture

The sewage serves as a breeding ground for mosquitoes and other disease-causing organisms. In many countries, sewage is treated and used for drinking and also for industrial and agricultural purposes, such as in culturing fish and field crops. Fish are known to convert bio-waste into edible protein efficiently. India, Germany, USSR, Poland, Hungary, Israel, China and Indonesia have obtained an annual fish production of 4-10 t/ha/yr by using domestic sewage which serves as a source of fertilizer and feed for fish ponds and fish respectively. Fish culture in sewage systems involves meager recurring expenses but the return is substantial. A fish yield of 3 t/ha/yr has been reported from oxidation ponds receiving effluent from septic tanks diluted with fresh water in the ratio of 1: 3.

Quality of Sewage

The water content of domestic sewage is about 99 per cent. Though the rest consists of organic and inorganic constituents, the former forms three-quarters of the total quantity. The common gases and salts present in sewage are carbon dioxide, hydrogen, sulphide, ammonia, nitrate and phosphate. The other heavy metals present in traces are zinc, copper, chromium, manganese, nickel and lead. It is worth mentioning here that the dissolved oxygen content of sewage is almost nil. Apart form the abiotic components, the sewage harbours a variety of bacteria and protozoans. However, the sewage is beneficial for fish culture because, (a) the smaller particles of sewage form direct food source to zooplankton and benthos, (b) the larger particles of sewage are directly utilised by fish, and (c) soluble organic substances are directly used by phytoplankton for photosynthesis and by zooplankton as a food source.

Sewage Treatment

Owing to lack of oxygen, the sewage is not suitable directly for fish culture. Hence before letting it out to fish ponds it is to be treated by any of the following methods:

1. Cascading sewage through a series of ponds;
2. Filtration of sewage in two stages followed by sedimentation;
3. Aeration of sewage; and
4. Dilution of sewage with fresh-water (one part sewage with five parts of fresh-water).

However, treatment of sewage through stabilization is easier and effective compared to other methods. In this method, the sewage is treated in the oxidation pond as follows.

The sewage is first stagnated in the oxidation pond for about 10 days to allow the particulate organic matter (coarse particles) to settle down. In the meantime, the dissolved organic matter is decomposed by aerobic microorganisms resulting in the formation of inorganic nutrients such as nitrate, phosphate, sulphate, etc. The particulate matter settled at the bottom, on the other hand, is partially decomposed into various organic acids, methane, hydrogen sulphide, ammonia, etc., owing to the limited availability of oxygen. As a consequence of the aerobic decomposition mentioned above, the nutrient status of the water is improved leading to the propagation of algae and increase in the dissolved oxygen level, which helps in the further oxidation of the partially decomposed organic substances into inorganic substances. As a result of these changes, the sewage in the oxidation pond becomes clear and is enriched with oxygen and nutrients and is, thus, made suitable for fish culture.

To remove the toxic heavy metals, it is advisablc to grow water hyacinth, *Eichhornia crassipes* at one corner of the oxidation pond.

Description of Oxidation Ponds

As far as possible, the oxidation pond which also serves as a storage for incoming sewage, should be made of concrete. Further, both the oxidation ponds and fish ponds should be located in open space, so that; the sewage is treated at a shorter time due to its contact with atmosphere. Alternatively, the sewage may be conveyed through sloppy open channels to oxidation ponds.

To make sewage-fish culture economical, the sewage may be conveyed to oxidation pond by gravitation. Arrangements could be made to discharge the sewage at the bottom of the oxidation pond, so that, there will be least disturbance in the sewage column, particularly in the supernatant portion, which is to be taken to the fish ponds again by gravitation. Considering the above, the oxidation pond should be located on higher elevation than the fish ponds.

The size of the oxidation pond could be so designed to discharge the sewage for 10 days continuously. For example, if the daily inflow of sewage is one lakh litres (100 m^3), the size of the oxidation ponds may be 50 × 20 × 1.5 m. Though 1 m depth may be enough for storing the sewage, 0.5 m free board or space above the water column is essential.

Sewage-Fed Fish Ponds

The number of fish ponds should be calculated in such a way that their total area is equal to the area of oxidation pond. All species suitable for fresh-water fish culture are also suitable for sewage-fed fish culture. However, the Indian major carps, *viz.* catla, rohu and mrigal are more suitable. Considering the rich nutrient status of the treated sewage and high survival rate (70-80 per cent) of fingerlings in sewage, the stocking density of carp fingerlings can be 20,000/ha (catla 5,000, rohu 10,000 and mrigal 5,000 in number). A total fish production of 6 t/ha has been achieved by sewage-fish culture. Alternatively, the Indian major carps, silver carp, grass carp, common carp and fringe-lipped carp may also be stocked in sewage-fish ponds at a stocking density of 10,000/ha as shown in Table 16.

It may be observed that such a fish yield results without the addition of fertilizers and supplemental feeds. Further, the growth rate of fish has also been found to be high in sewage systems. It is reported recently, that high yield of giant fresh-water prawn, *Macrobrachium rosenbergii* could be obtained from rice fields fertilized with domestic sewage.

Table 16: Stocking Ratio for Sewage-Fed Fish Ponds

Variety of Fish	*6 Variety Stocking (per cent)*	*5 Variety Stocking (per cent)*
Silver carp	25	–
Catla	15	40
Rohu	10	10
Mrigal	25	20
Common carp	20	20
Grass carp	5	–
Fringe-lipped carp	–	10

Air-breathing and low-oxygen-tolerating fish like murrels and catfish may also be grown along with the species already mentioned; however, they should be introduced only after the fingerlings of other fish have grown to a considerable size. In oxidation ponds, murrels can be directly cultured along with the rapid breeding tilapia, the young of which form food for murrels.

Aesthetic considerations may often limit the use of sewage-cultured fish. These can be overcome by transferring the harvested live-fish to fresh-water ponds and keeping them for a couple of weeks when the objectionable odours and pathogens, if any, will be removed making the fish marketable (Depuration). Bacteriological tests conducted on the skin, gut and muscle of fishes grown in sewage and fresh-water showed that species of *Salmonella* were not present in any of the fishes. *Shigella* was isolated from fishes reared in sewage-fed ponds. But neither *Salmonella* nor *Shigella* was present in cooked fish.

Sewage Water for Other Crops

The drained sewage could be used for irrigation, especially for fodder crops, vegetable crops, coconut, palms, etc.

Economics of Sewage-Fed Fish Culture

The recurring expenditure on sewage-fed fish culture is meager compared to that of fresh-water fish culture. The economics of sewage-fed fish culture shows that this type of culture is lucrative and a fish farmer can obtain an income, on an average, of Rs. 41,000/ha/yr. If murrels are cultured in oxidation ponds and the excess sewage is utilized for the cultivation of crops, the revenue could be further augmented.

References

Ahilan, B., Ravaneshwaran, K. and Kumaraval, P., 2011. Integrated Aquaculture. Daya Publishing House, New Delhi, 142 pp.

Agarwal, S.C., 1994. A Handbook of Fish Farming. Narendra Publication House, Delhi, 117 pp.

Ayyappan, S., Jena, J.K., Gobalakrishnan, A. and Pandey, A.K., 2006. Handbook of Fisheries and Aquaculture, ICAR, New Delhi, 755 pp.

Boyd, C.E., 1995. Bottom soils, sediment and pond aquaculture. Chapman and Hall, New York, 348pp.

David, C. and James, M. 1987. A Guide to Integrated warm water Aquaculture. Institute of Aquaculture Publications, University of Stirling, Scotland. 238pp.

Hora, S.L. and Pillay, T.V.R., 1962. Handbook on fish culture in the Indo-pacific region. FAO Fish. Tech. Paper 14. 204 pp.

Huet, Marcel, 1994. Textbook of Fish Culture. University Press, Cambridge, 437 pp.

Jhingran, V.G., 1991. Fish and Fisheries of India. Hindustan Publishing Corporation, Delhi, 622 pp.

Lockeretz., W., 2007. Organic Farming. Cromwell Press, Trowbridge, 282 pp.

Pillay, T.V.R., 1993. Aquaculture Principles and Practices. Fishing News Books, England, 515 pp.

Pillay, T.V.R. and Dill, Wm.A., 1976. Advances on Aquaculture. Fishing News Books, Farnham. Surrey, England, 653 pp.

Rath, R.K., 1993. Freshwater Aquaculture. Scientific Publishers, Jodhpur, 193 pp.

Santhanam, R., Sukumaran, N. and Natarajan, P., 1990. A Manual of Freshwater Aquaculture. Oxford and IBH Publishing Co. Pvt. Ltd., New Delhi, 193 pp.

Santhanakumar, G. and Selvaraj, A.M., 1995. Concepts of Aquaculture. Meenam Publications, Nagercoil, 204 pp.

Stickney, R., 1979. Principles of Warm Water Aquaculture. John Wiley and Sons, USA, 375 pp.

Stickney, R., 1994. Principles of Aquaculture. John Wiley and Sons, USA, 502 pp.

Index